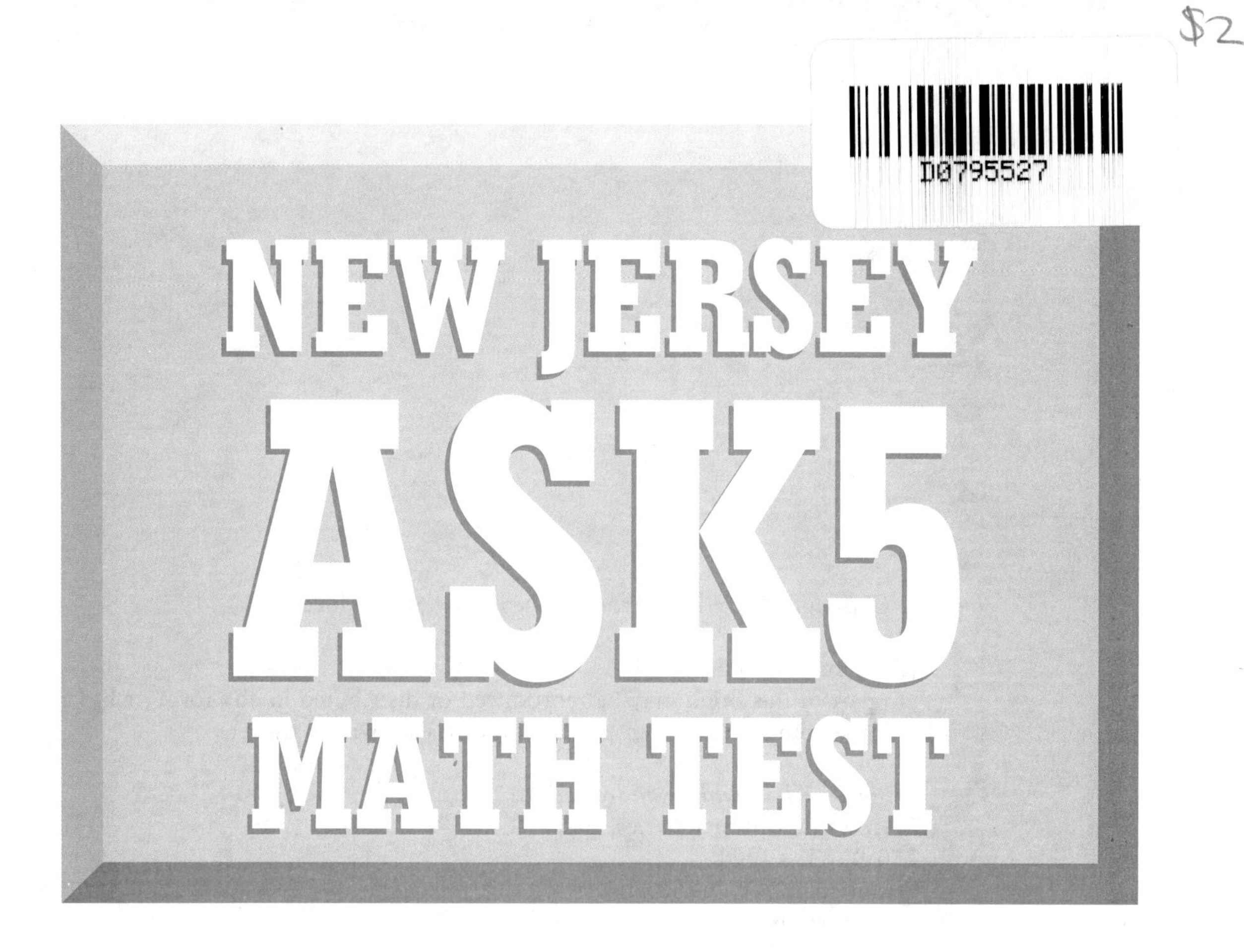

Eric Milou, Ed.D.
Professor of Mathematics, Rowan University, Glassboro, NJ

Melissa Jackson
Mathematics Teacher, Monongahela Middle School, Deptford, NJ

All inquiries should be addressed to:
Barron's Educational Series, Inc.
250 Wireless Blvd.
Hauppauge, NY 11788
www.barronseduc.com

ISBN-13: 978-0-7641-4238-3
ISBN-10: 0-7641-4238-0

Library of Congress Catalog Card No.: 2008055884

Library of Congress Cataloging-in-Publication Data

Milou, Eric.
New Jersey ASK5 math test / Eric Milou, Melissa Jackson.
p. cm.
Includes index.
ISBN-13: 978-0-7641-4238-3
ISBN-10: 0-7641-4238-0
1. New Jersey Assessment of Skills and Knowledge—Study guides. 2. Mathematics—Examinations—New Jersey—Study guides. 3. Mathematics—Study and teaching (Elementary)—New Jersey. 4. Fifth grade (Education)—New Jersey. I. Jackson, Melissa. II. Title. III. Title: New Jersey ASK 5 math test.

QA43.M516 2009
510.76–dc22

2008055884

Printed in the United States of America
9 8 7 6 5 4 3 2

CONTENTS

New Jersey Assessment of Skills and Knowledge
Grade 5
MATHEMATICS REFERENCE SHEET

Use the information below to answer questions on the Mathematics section of the Grade Five Assessment of Skills and Knowledge (NJ ASK 5).

The sum of the measures of the interior angles of a triangle = 180°

Distance = rate × time

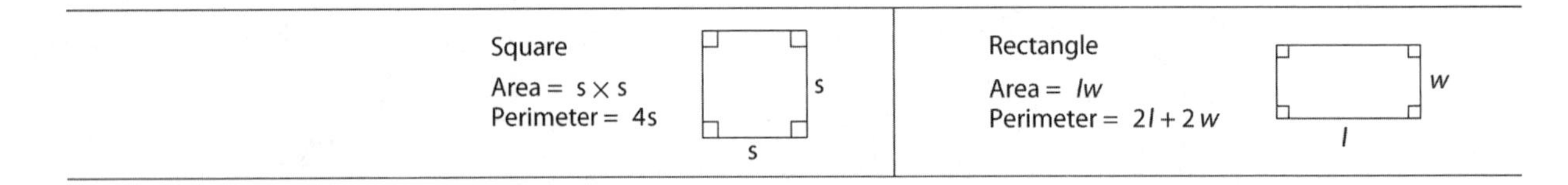

USE THE FOLLOWING EQUIVALENTS FOR YOUR CALCULATIONS

60 seconds = 1 minute 60 minutes = 1 hour 24 hours = 1 day 7 days = 1 week 12 months = 1 year 365 days = 1 year 52 weeks = 1 year	12 inches = 1 foot 3 feet = 1 yard 36 inches = 1 yard 5,280 feet = 1 mile 1,760 yards = 1 mile	10 millimeters = 1 centimeter 100 centimeters = 1 meter 10 decimeters = 1 meter 1000 meter s = 1 kilometer
8 fluid ounces = 1 cup 2 cups = 1 pint 2 pints = 1 quart 4 quarts = 1 gallon 1000 milliliters (mL) = 1 liter (L)	16 ounces = 1 pound 2,000 pounds = 1 ton 1000 milligrams = 1 gram 100 centigrams = 1 gram 10 grams = 1 dekagram 1000 grams = 1 kilogram	

Source: http://www.state.nj.us/education/assessment/ms/5-7/gr5_NJASK06_math_ref.pdf

Mathematics Manipulatives Sheet
Shapes, Protractor, and Ruler
Grades 5, 6, and 7 Only*

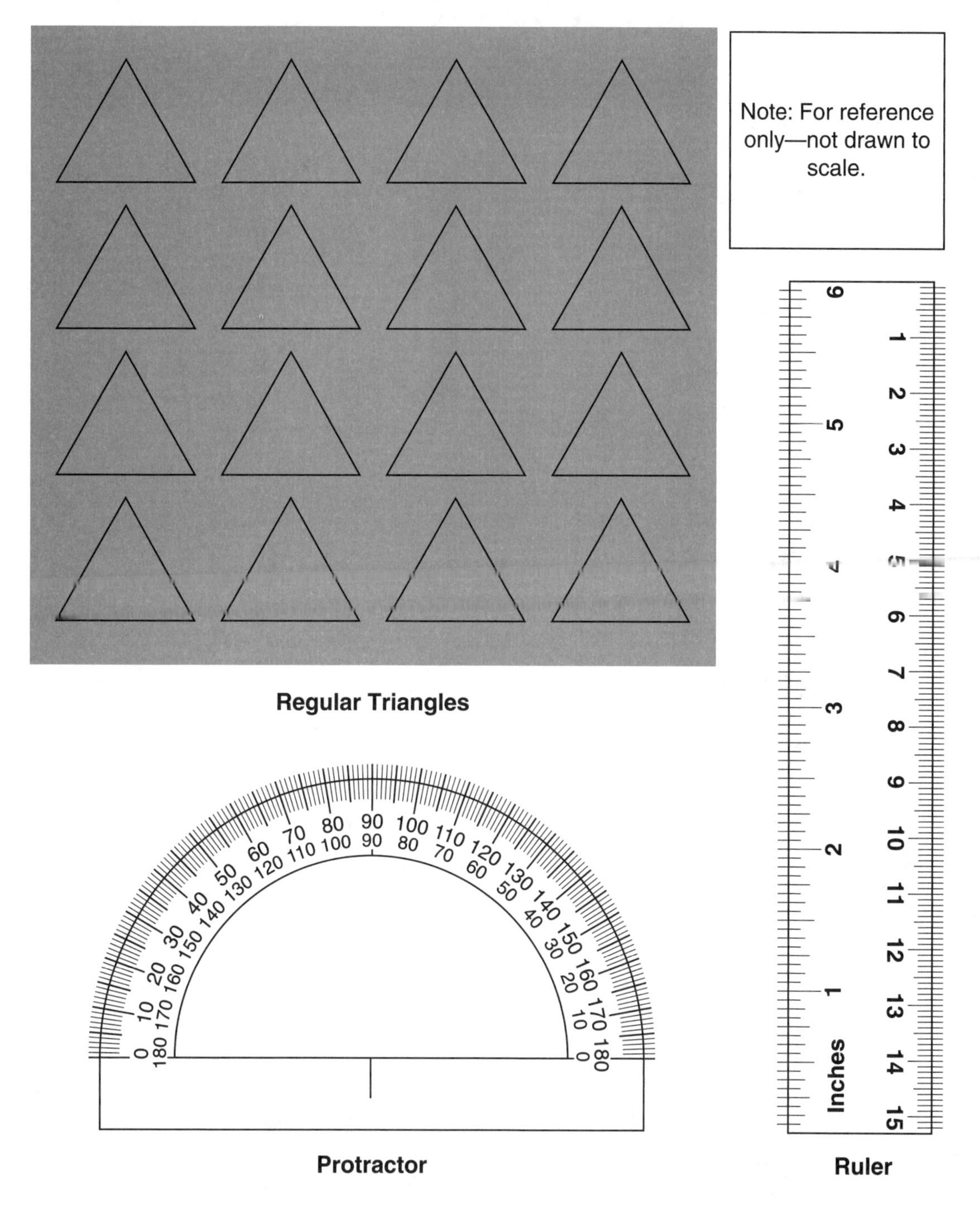

9-13330

Source: http://www.state.nj.us/education/assessment/ms/5-7//2006-07ManipulativesSheet.pdf

Chapter 1

INTRODUCTION TO THE TEST

The enactment of the No Child Left Behind Act of 2001 (NCLB) required that each state administer annual standards-based assessments to students in grades 3 through 8, and at least once in high school. Federal expectations are that each state will provide tests that are grounded in the state's content standards and that assess students' thinking skills. For more information, see *http://www.nj.gov/education/assessment/es/*.

In response to NCLB requirements, New Jersey required all students to take the *The New Jersey Assessment of Skills and Knowledge (NJASK)* in grades 3–8. The purpose of the NJASK is to measure the level of mathematics proficiency that New Jersey students have achieved by spring of each grade. Each assessment measures the NJ Mathematics Standards at the respective grade level as follows:

1. Number and Numerical Operations
   a. Number Sense
   b. Numerical Operations
   c. Estimation
2. Geometry and Measurement
   a. Geometric Properties
   b. Transforming Shapes
   c. Coordinate Geometry
   d. Units of Measurement
   e. Measuring Geometric Objects

3. Patterns and Algebra
   a. Patterns
   b. Functions and Relationships
   c. Modeling
   d. Procedures

4. Data Analysis, Probability, and Discrete Mathematics
   a. Data Analysis (Statistics)
   b. Probability
   c. Discrete Mathematics—Systematic Listing and Counting
   d. Discrete Mathematics—Vertex-Edge Graphs and Algorithms

This book is composed of chapters addressing each of the NJASK5 expectations.

NJASK5 TEST DESIGN

The NJASK5 contains three types of questions including multiple choice (MC, 1 point each), short constructed response (SCR, 1 point each), and extended constructed response (ECR, maximum of 3 points each). The table below displays the total number of questions and the item count by question type.

		Grade 5
Item count by type	MC	35
	SCR	6
	ECR	3
# of sections		5
Total raw score points possible (Excludes field-tested items)		50
Total testing time over 2 days		114

The MC items are 1 point each (no partial credit). The SCR items are similar to MC items: students must write their actual answer in the space provided; and no partial credit is given. The answer is either right (1 point) or wrong (0 points). The ECR items are graded on a 0 to 3 point rubric. ECR items will require responses that range in length from several numbers to three to four steps. Moreover, ECR items may ask for a graph (or figure or diagram or table) with labels or a number sentence to support the figure. ECR items may also require a more complex solution process including communicating the mathematical ideas or results. Scoring rubrics for ECR items will focus on conceptual understanding, application of appropriate procedures, and accuracy. ECR items will be scored with a four-level scoring rubric (0–3 points). The NJ holistic scoring guide (generic rubric) for scoring ECR items is shown below.

3-Point Response: The response shows complete understanding of the problem's essential mathematical concepts. The student executes procedures completely and gives relevant responses to all parts of the task. The response contains few minor errors, if any. The response contains a clear, effective explanation detailing how the problem was solved so that the reader does not need to infer how and why decisions were made.

2-Point Response: The response shows nearly complete understanding of the problem's essential mathematical concepts. The student executes nearly all procedures and gives relevant responses to most parts of the task. The response may have minor errors. The explanation detailing how the problem was solved may not be clear, causing the reader to make some inferences.

1-Point Response: The response shows limited understanding of the problem's essential mathematical concepts. The response and procedures may be incomplete and/or may contain major errors. An incomplete explanation of how the problem was solved may contribute to questions as to how and why decisions were made.

0-Point Response: The response shows insufficient understanding of the problem's essential mathematical concepts. The procedures, if any, contain major errors. There may be no explanation of the solution or the reader may not be able to understand the explanation. The reader may not be able to understand how and why decisions were made.

Problem-specific rubrics are given for sample ECR items in each chapter of this book.

TEST ADMINISTRATION

The NJASK5 is given over a 2-day period in early May as follows:

- Day #1 is composed of

Section #1, 6 SCR, 12 minutes, **no calculator**

Section #2, 12 MC, 1 ECR, 30 minutes, calculator allowed

- Day #2 is composed of

Section #3, 12 MC, 1 ECR, 27 minutes, calculator allowed

Section #4, 11 MC, 1 ECR, 25 minutes, calculator allowed

Section #5, 10 MC, 1 ECR, 23 minutes, calculator allowed

Note that one section on day #2 is composed of field-tested items that do not count in the total raw score possible.

PASSING SCORE

The state of NJ grades the NJASK5 on a scale of 100 to 300. A 200 score is proficient (passing), a 250 score is advanced proficient. As of September 2008, the proficient score (200) on the NJASK5 is 50% correct. The advanced proficient score (250) is 80% correct.

ANSWER SHEET: DIAGNOSTIC TEST

Section 1
Short Constructed-Response Questions

1. ______________________________

2. ______________________________

3. ______________________________

4. ______________________________

5. ______________________________

6. ______________________________

Section 2
Multiple-Choice Questions

7. Ⓐ Ⓑ Ⓒ Ⓓ
8. Ⓐ Ⓑ Ⓒ Ⓓ
9. Ⓐ Ⓑ Ⓒ Ⓓ
10. Ⓐ Ⓑ Ⓒ Ⓓ
11. Ⓐ Ⓑ Ⓒ Ⓓ
12. Ⓐ Ⓑ Ⓒ Ⓓ
13. Ⓐ Ⓑ Ⓒ Ⓓ
14. Ⓐ Ⓑ Ⓒ Ⓓ
15. Ⓐ Ⓑ Ⓒ Ⓓ
16. Ⓐ Ⓑ Ⓒ Ⓓ
17. Ⓐ Ⓑ Ⓒ Ⓓ
18. Ⓐ Ⓑ Ⓒ Ⓓ

Extended Constructed-Response Questions

19.

Section 3
Multiple-Choice Questions

20. Ⓐ Ⓑ Ⓒ Ⓓ
21. Ⓐ Ⓑ Ⓒ Ⓓ
22. Ⓐ Ⓑ Ⓒ Ⓓ
23. Ⓐ Ⓑ Ⓒ Ⓓ
24. Ⓐ Ⓑ Ⓒ Ⓓ
25. Ⓐ Ⓑ Ⓒ Ⓓ
26. Ⓐ Ⓑ Ⓒ Ⓓ
27. Ⓐ Ⓑ Ⓒ Ⓓ
28. Ⓐ Ⓑ Ⓒ Ⓓ
29. Ⓐ Ⓑ Ⓒ Ⓓ
30. Ⓐ Ⓑ Ⓒ Ⓓ
31. Ⓐ Ⓑ Ⓒ Ⓓ

Extended Constructed-Response Questions

32.

Section 4
Multiple-Choice Questions

33. Ⓐ Ⓑ Ⓒ Ⓓ
34. Ⓐ Ⓑ Ⓒ Ⓓ
35. Ⓐ Ⓑ Ⓒ Ⓓ
36. Ⓐ Ⓑ Ⓒ Ⓓ
37. Ⓐ Ⓑ Ⓒ Ⓓ
38. Ⓐ Ⓑ Ⓒ Ⓓ
39. Ⓐ Ⓑ Ⓒ Ⓓ
40. Ⓐ Ⓑ Ⓒ Ⓓ
41. Ⓐ Ⓑ Ⓒ Ⓓ
42. Ⓐ Ⓑ Ⓒ Ⓓ
43. Ⓐ Ⓑ Ⓒ Ⓓ

Extended Constructed-Response Question

44.

DIAGNOSTIC TEST

SECTION 1

SHORT CONSTRUCTED RESPONSE

Directions: For each question, write the correct answer on the line provided on the answer sheet. You may NOT use a calculator for this section.

1. A gallon contains 128 ounces. Paul wants to divide 3 gallons of apple cider equally among the 2 dozen friends at his party. How many ounces of apple cider will each friend receive?

2. What value of B would make the following number sentence true? $10 - B = 7$

3. On five tests of 100 points each, Tamika has an average of exactly 90. What is the lowest score she could have made on any of the five tests?

4. Eric has a number cube with the number 1 on five of the six faces. What is the probability of NOT rolling a 1 on his next roll?

5. If there are ten marbles in a bag, five red and five green, what is the probability that a marble picked from the bag will be red?

6. Examine the equation: $25 \times A = 175$. What value of A makes this equation correct?

SECTION 2

MULTIPLE CHOICE

Directions: Darken the letter of the best answer on the answer sheet. You may use a calculator for this section.

7. Richardo has a 20-gallon fish tank. Which metric unit of measure is best for calculating the volume of the tank?

   A. liter

   B. centimeter

   C. millimeter

   D. meter

8. Jim bought 5 pieces of wood. They measured 13.25 inches, 13.3 inches, 13.008 inches, 12.999 inches, and 13.03 inches in length. Which of the choices below shows the pieces in order from shortest to longest?

   A. 12.999 in., 13.25 in., 13.3 in., 13.008 in., 13.03 in.

   B. 12.999 in., 13.3 in., 13.008 in., 13.25 in., 13.03 in.

   C. 12.999 in., 13.008 in., 13.25 in., 13.03 in., 13.3 in.

   D. 12.999 in., 13.008 in., 13.03 in., 13.25 in., 13.3 in.

9. The students want to bake 200 cupcakes to sell at the school's bake fair. They have baked 70 so far. Jaime wants to write an equation to find how many more cupcakes, *c*, need to be baked. Which equation should Jaime write?

   A. $c \times 70 = 200$

   B. $c - 70 = 200$

   C. $70 - c = 200$

   D. $70 + c = 200$

10. Examine the shapes below. Which one has a right angle?

A. The pentagon

B. The octagon

C. The triangle

D. None of them

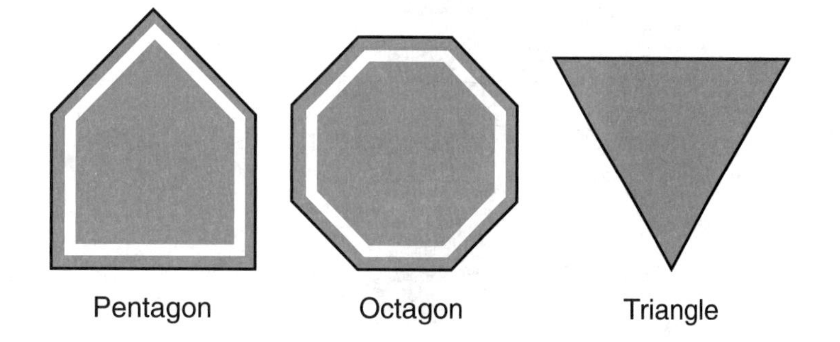

11. Use estimation on the following: 21,100 – 19,078 is approximately _____.

A. a little more than 1,000

B. a little more than 1,500

C. a little more than 2,000

D. a little more than 2,500

12. Melissa drew a line segment that was 1 inch long. Which is closest to 1 inch?

A. 2.5 centimeters

B. 25 centimeters

C. 2.5 meters

D. 25 meters

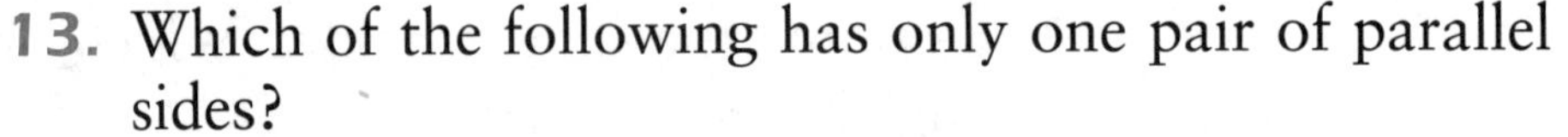

13. Which of the following has only one pair of parallel sides?

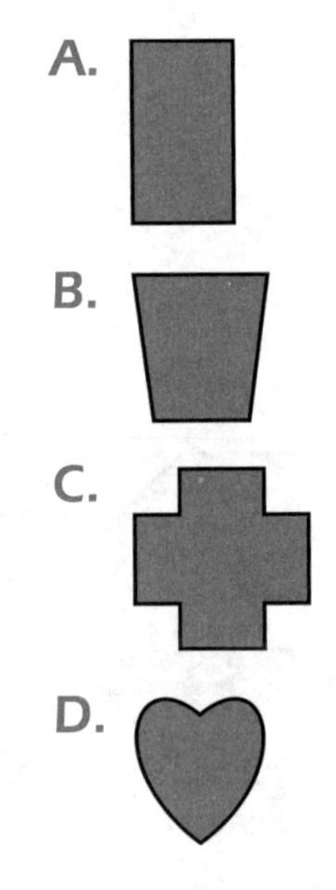

14. Which best describes the figure below?

A. Line segment *AB*

B. Ray *AB*

C. Line *AB*

D. Radius *AB*

15. Consider the relationship between the radius and the diameter of a circle. Which is a true statement?

A. The length of the radius of a circle is equal to the length of the diameter.

B. The length of the radius of a circle is always greater than the length of the diameter.

C. The length of the radius of a circle is one-half the length of the diameter.

D. The length of the radius of a circle is four times the length of the diameter.

16. Consider the following pattern: 1, 5, 25, 125. If this pattern of numbers continues, what is the next number in the pattern?

A. 150

B. 225

C. 625

D. 1,250

17. In Mr. Heinz's class, there are 8 students with brown hair, 5 students with blond hair, and 12 students with black hair. One student is selected to answer a problem on the board, what is the probability that the student selected has black hair?

A. 8/25

B. 5/25

C. 12/25

D. 13/25

18. On Saturday, it snowed two and one-half inches. Which of the following shows that amount?

A. 25

B. 2.5

C. 0.25

D. 2.05

EXTENDED CONSTRUCTED-RESPONSE QUESTION

Directions: Write your response in the space provided on the answer sheet. Answer the question as completely as possible. You may use a calculator for this section.

19. The Dueling Dragons Roller Coaster consists of two roller coasters, Fire and Ice. During the ride these roller coasters narrowly pass within inches of each other. Each roller coaster carries 32 people for a ride time of approximately 4 minutes, which includes the load time.

- What is the maximum number of 4-minute roller coaster rides there will be in one hour? Show all work and explain answers.
- If each roller coaster ride was completely full, how many people would ride the Dueling Dragons Roller Coaster in 1 hour? Show all work and explain answers.

SECTION 3

MULTIPLE CHOICE

Directions: Darken the letter of the best answer on the answer sheet. You may use a calculator for this section.

20. A movie started at 6:30 p.m. and lasted for 1 hour and 45 minutes. At what time did the movie end?

A. 7:45 p.m.

B. 8:00 p.m.

C. 8:15 p.m.

D. 8:30 p.m.

21. Which of the following is not a parallelogram?

A. Trapezoid

B. Square

C. Rectangle

D. Rhombus

22. Mark's desk at school has a length of 3 ft and a width of 4 ft. What is the perimeter of Mark's desk at school?

A. 12 square feet

B. 12 feet

C. 14 square feet

D. 14 feet

23. A large pizza costs $9 at Al's Pizza Shop. If Al sold 54 pizzas, how much money did he collect?

A. $45

B. $63

C. $400

D. $486

24. The table below shows the average temperature in each month in NJ during the first six days of the year. What was the mean temperature during the first six days of January?

Date	Temperature (°F)
Jan 1	30
Jan 2	35
Jan 3	20
Jan 4	10
Jan 5	5
Jan 6	20

A. 20

B. 21

C. 25

D. 30

25. Which of the shapes below could never have perpendicular lines?

A. Circle

B. Square

C. Rectangle

D. Triangle

26. Find the rule that describes the following pattern: 2, 4, 3, 5, 4, 6, 5, . . .

A. Multiply by 2, subtract 1

B. Add 2, subtract 1

C. Multiply by 2, add 1

D. Add 2, subtract 2

27. According to a report published in 2006, the population of NJ was 8,724,560. What does the 7 in this number represent?

A. Seven million

B. Seven hundred thousand

C. Seven thousand

D. Seven hundred

28. Which is a prime factor of the composite number 24?

A. 1

B. 2

C. 4

D. 6

29. Stephanie bought a bag of 24 apples. Her family ate $\frac{3}{8}$ of the bag. How many apples remained?

A. 5

B. 9

C. 15

D. 21

30. If the pattern below continues, what is the next number? 3, 5, 6, 8, 9, 11, 12

A. 13

B. 14

C. 15

D. 16

31. A measuring cup has a capacity of $\frac{3}{4}$ of a cup. How many times must it be filled to result in 3 cups?

A. 2 times

B. 3 times

C. 4 times

D. 5 times

EXTENDED CONSTRUCTED-RESPONSE QUESTION

Directions: Write your response in the space provided on the answer sheet. Answer the question as completely as possible. You may use a calculator for this section.

32. The township requires that pools be fenced in as a safety measure. Rick purchased 24 linear yards of fencing and is trying out different dimensions to determine the most appropriate for his pool.

- Use graph paper to draw all possible rectangular fencing dimensions that are possible with 24 linear yards of fencing.
- Determine which of the fencing dimensions would create the greatest amount of area for the pool.

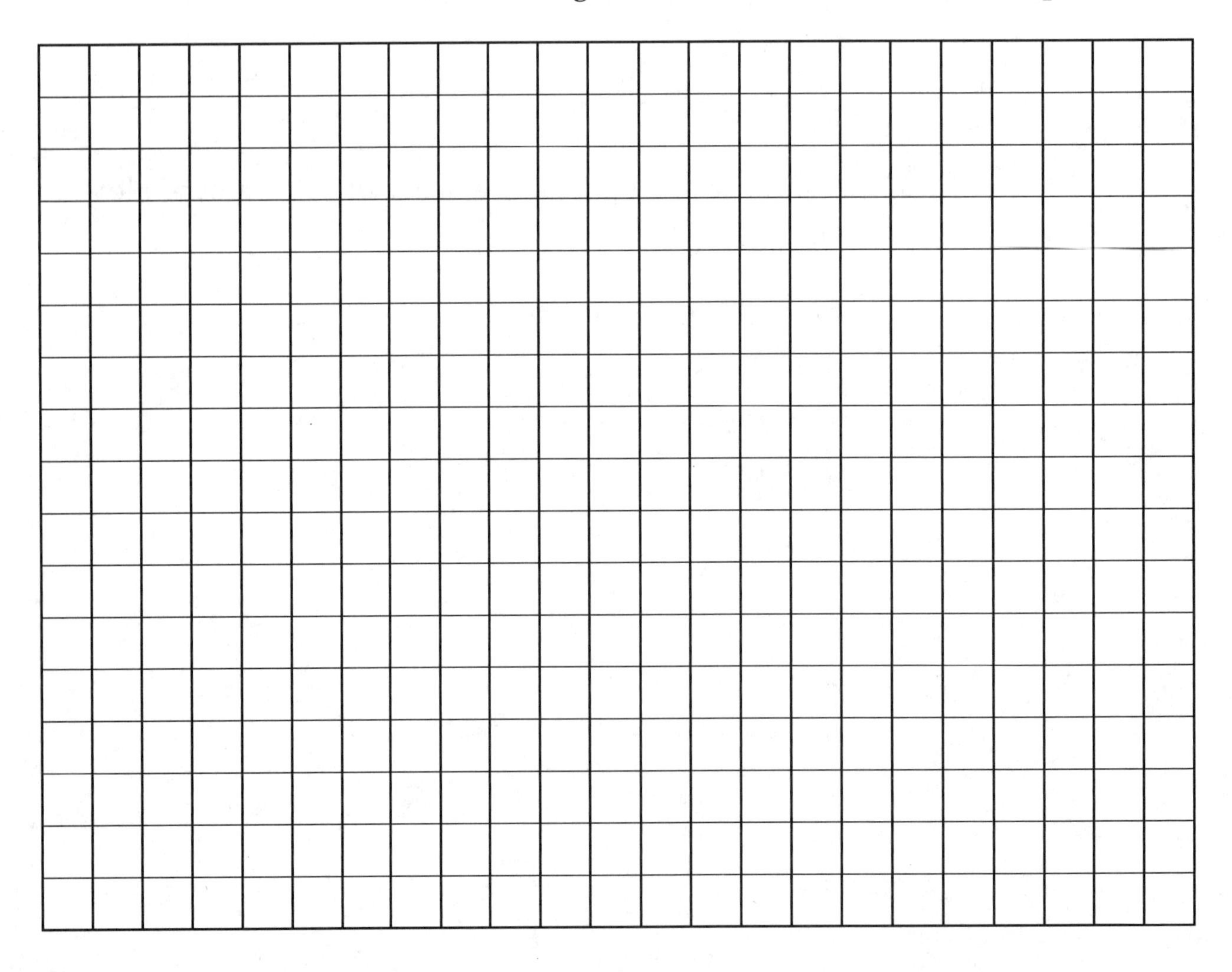

SECTION 4

MULTIPLE CHOICE

Directions: Darken the letter of the best answer on the answer sheet. You may use a calculator for this section.

33. The lawn company was mowing a large football field. They have completed mowing $\frac{2}{3}$ of the yard when it was time to break for lunch. What fraction is equivalent to $\frac{2}{3}$?

A. $\frac{3}{4}$

B. $\frac{4}{6}$

C. $\frac{3}{6}$

D. $\frac{3}{2}$

34. Given the input-output table below, which of the statements correctly describes the relationship between the IN and OUT number?

IN	OUT
2	5
3	7
4	9
5	11

A. OUT = IN × 2

B. OUT = IN + 2

C. OUT = IN × 2 + 1

D. OUT = (IN + 1) × 2

35. Using the spinner below, what is the probability that the spinner will land on a prime number?

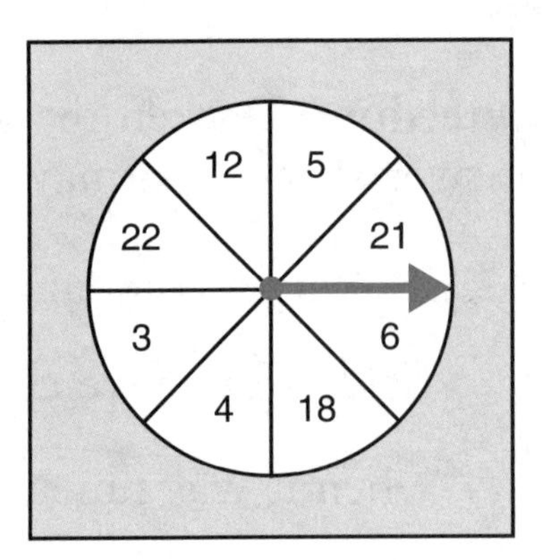

A. $\frac{1}{8}$

B. $\frac{1}{4}$

C. $\frac{1}{3}$

D. $\frac{1}{2}$

36. Perry is making punch for the fifth grade party. He uses 1 pint of mix, 4 quarts of water, and 6 cups of fruit. His mixture will serve 20 students. There are 100 students in the fifth grade. How many quarts of water will Perry need to serve all 100 students?

A. 4

B. 8

C. 16

D. 20

37. Jason collects baseball cards every week. He collects 20 cards per week for 7 weeks. Which expression can be used to find the number of cards that Jason collects?

A. 20×7

B. $20 \div 7$

C. $20 + 7$

D. $20 - 7$

38. The angle below was draw by Mrs. Terrill. What kind of angle is it?

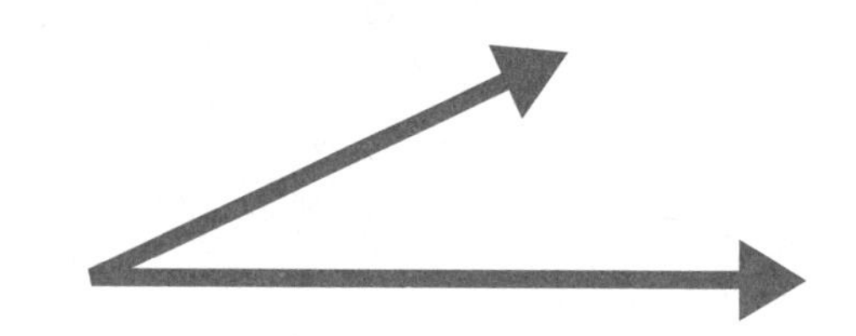

A. Acute

B. Obtuse

C. Right

D. Large

39. The perimeter of the quadrilateral below is 100 inches. What is the length of the missing side?

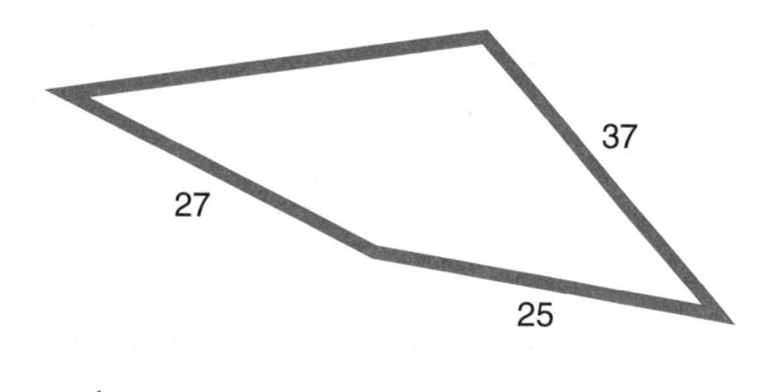

A. 27 inches

B. 20 inches

C. 15 inches

D. 11 inches

40. Rosemary examines eight different bags of candies. The amount of candy in each bag was 54, 53, 52, 54, 55, 56, 52, and 52 pieces. What is the mode of her data set?

A. 52

B. 53

C. 54

D. 57

41. A normal penny was flipped ten times and each time it landed on tails. What is the probability that on the next flip of the coin, the penny will land on tails?

A. 0

B. $\frac{1}{2}$

C. $\frac{9}{10}$

D. 1

42. Norma's Ice Cream store has the following treats:

Ice Cream Flavors	Toppings	Sauces
Vanilla	Jimmies	Hot fudge
Chocolate	Nuts	Butterscotch
Strawberry	Whipped cream	

How many different types of cones can be made?

A. 8

B. 9

C. 11

D. 18

43. Estimate 59×9. Which pair of numbers is the product between?

A. 60 and 100

B. 200 and 400

C. 300 and 500

D. 500 and 700

EXTENDED CONSTRUCTED-RESPONSE QUESTION

Directions: Write your response in the space provided on the answer sheet. Answer the question as completely as possible. You may use a calculator for this section.

44. Michele took a 20-question quiz in her mathematics class. She answered 12 of the questions correctly on the quiz.

- What is Michele's score written as a percent? Explain how you found the percent.
- Michele took a 25-question quiz in her science class. She answered 16 of the questions correctly. Is the score on her science quiz the same or different than her math score? Explain why your answer is correct.

ANSWERS TO DIAGNOSTIC TEST

1. $(3 \times 128) = 384/24 = 16$ ounces

2. $B = 3$

3. $90 \times 5 = 450$. If she scores 100 on four of the tests, the lowest score possible on the fifth test would be 50.

4. $\frac{1}{6}$

5. $\frac{5}{10} = \frac{1}{2}$

6. $25 \times 7 = 175$

7. A. liters

8. D. 12.999 in., 13.008 in., 13.03 in., 13.25 in., 13.3 in.

9. D. $70 + c = 200$

10. A. The pentagon

11. C. $21{,}100 - 19{,}078 = 21{,}000 - 19{,}000 = 2{,}000$

12. A. 2.5 centimeters is approximately 1 inch.

13. B. A and C have more than one pair of parallel sides, and D has none.

14. C. Line *AB*

15. C. The length of the radius of a circle is one-half the length of the diameter.

16. C. 625. Each number is 5 times the previous one.

17. C. 12/25

18. B. 2.5

19. You will receive full credit if:

    - You correctly find that there are 15 four-minute rides for each of the roller coasters in a 1-hour period. You may also find that there are 30 rides total for both roller coasters.
    - You correctly find that 64 people times 15 rides or 960 people would ride the roller coasters in 1 hour. You may also find the 960 by multiplying 32 people by 30 rides.
    - You show work and explain how you got your answers.

20. C. 6:30 + 1 hr = 7:30 + 45 minutes = 8:15 p.m.

21. A. Trapezoid is not a parallelogram as it only has one pair of parallel sides.

22. D. 14 feet = 3(2) + 4(2)

23. D. $9 × 54 pizzas = $486

24. A. (30 + 35 + 20 + 10 + 5 + 20) / 6 = 20

25. A. Circle

26. B. Add 2, subtract 1

27. B. Seven hundred thousand

28. B. $24 = 2 \times 2 \times 2 \times 3$

29. C. 15. (24) × (3/8) = 9 apples were eaten by the family, so 24 – 9 = 15 remained.

30. B. 14 add 2, add 1

31. C. $\frac{3}{4} \times 4 = 3$

32. You will receive full credit if:

- You correctly draw all possible dimensions of the pool: 1 yd × 11 yd, 2 yd × 10 yd, 3 yd × 9 yd, 4 yd × 8 yd, 5 yd × 7 yd, and 6 yd × 6 yd.
- You correctly show the area of the above dimensions.
- You correctly determine the greatest area for the pool using 24 yards of fencing is 6 yards by 6 yards or 36 square yards.

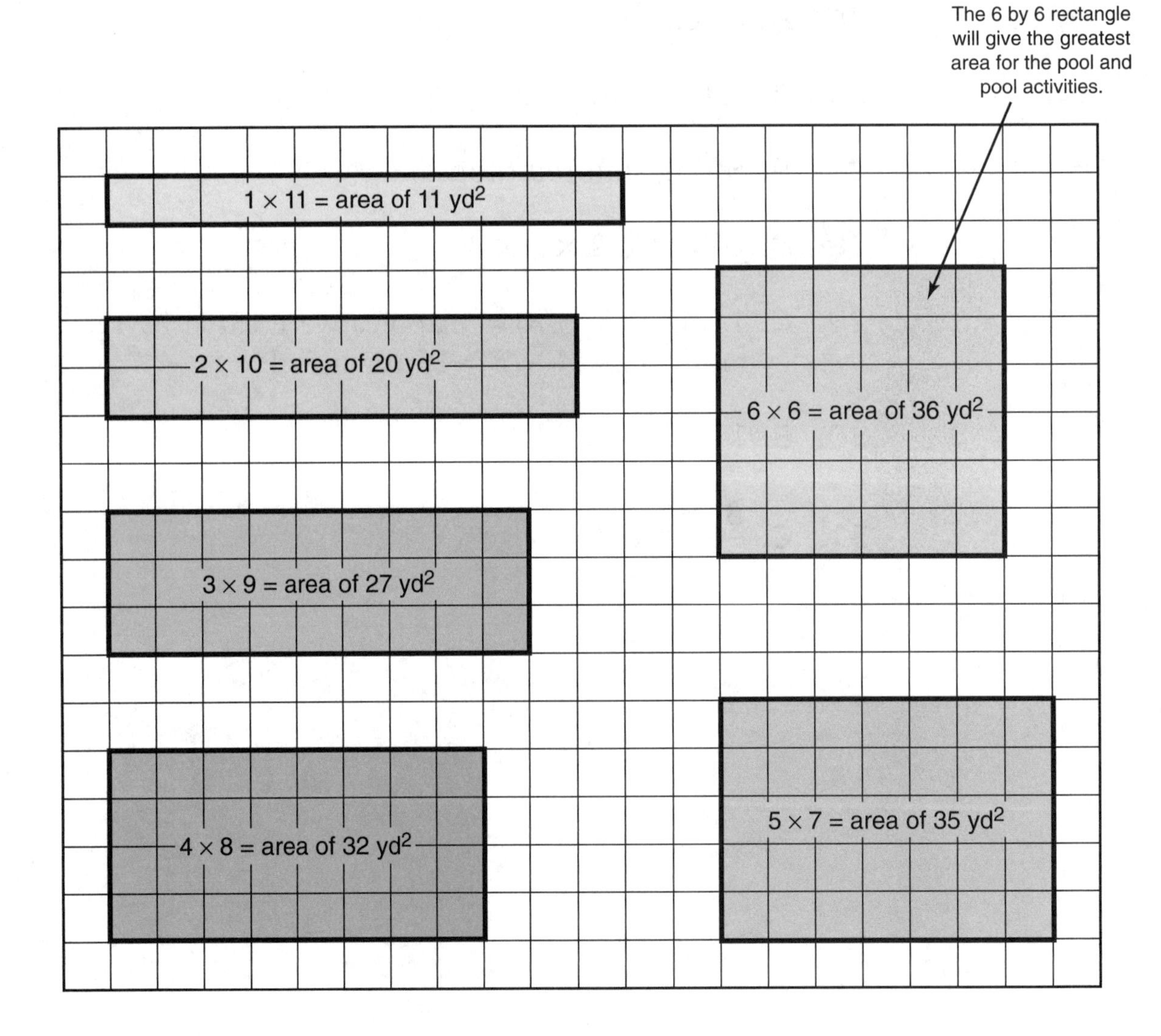

33. **B.** $\frac{4}{6} = \frac{2}{3}$

34. **C.** OUT = IN × 2 + 1

35. **B.** $\frac{2}{8} = \frac{1}{4}$. Only 2 of the numbers are prime (3 and 5).

36. **D.** 20. He needs to make 5 batches (100/20). 5 × 4 = 20 quarts.

37. **A.** 20 × 7

38. **A.** Acute—it is less than 90 degrees.

39. **D.** 100 – (27 + 25 + 37) = 11

40. **A.** 52 appears three times (most frequent = mode)

41. **B.** $\frac{1}{2}$ (probability remains the same)

42. **D.** 3 × 3 × 2 = 18

43. **D.** 500 and 700 (59 × 9 is approx 60 × 10 = 600)

44. You will receive full credit if:

- You correctly find 12/20 = 60%.
- You correctly find 16/25 = 64% and the science quiz is higher than the math quiz.

Chapter 3

NUMBERS AND NUMERICAL OPERATIONS

Numbers and arithmetic operations are what most of the general public think about when they think of mathematics. Numbers and operations remain at the heart of mathematical teaching and learning. The ability to choose the appropriate types of numbers and the appropriate operations for a given situation, and the ability to perform those operations as well as to estimate their results are all skills essential for modern day life.

In grade 5, test items refer to the following sets of numbers:

- Whole numbers through millions
- All fractions as part of a whole, as subset of a set, as a location on a number line, and as [divisions] or a representation for division
- All decimals

The Numbers and Numerical Operations strand includes three parts: Number Sense, Numerical Operations, and Estimation.

Number sense

- Is a feel for numbers and a common sense approach to using them.
- Is an ease with what numbers represent that comes from investigating their characteristics and using them in diverse situations.
- Involves an understanding of how different types of numbers, such as fractions and decimals, are related to each other, and how each can best be used to describe a particular situation.

- It includes the important concepts of place value, number base, magnitude, approximation, and estimation.

Numerical Operations include

- Using mental math.
- Using pencil-and-paper techniques.
- Using calculators.
- Understanding how to add, subtract, multiply, and divide whole numbers, fractions, decimals, and other kinds of numbers.

Estimation is a process that is used constantly in the real world. It is important that you are aware of the many situations in which an approximate answer is as good as, or even preferable to, an exact one.

Test items in the number strand may include

- Illustrations of the following: thousands blocks, hundreds blocks, tens blocks, ones blocks; unifix cubes, sticks, number lines; other counting manipulatives.
- Illustrations of real-life objects or simple geometric shapes, such as circles or rectangles for fractions or fractional parts.
- Number lines, which may be used when asking you to order and compare the magnitude of whole numbers and/or fractions (e.g., ordering of fractions on a number line should be limited to fractions with the same denominator).
- The symbols =, <, and > to compare whole numbers to whole numbers, fractions to fractions, and whole numbers to fractions.
- Pictures in extended-response items where you are asked to mathematically model what you see in the illustrations.
- Computation items that appear in either a vertical or horizontal format.

PLACE VALUE

Numbers, such as 123, 456, 789 have digits and each digit is a different place value. The chart below shows the value of each of the digits.

Hundred Millions	Ten Millions	Millions	Hundred Thousands	Tens Thousands	Thousands	Hundreds	Tens	Ones
1	2	3	4	5	6	7	8	9

PLACE VALUE EXERCISES

Write the following numbers.

1. 4 millions 6 hundred thousands 9 ten thousands 1 thousand 1 hundred 3 tens 1 one
2. 900,000 + 10,000 + 2,000 + 400 + 30 + 3
3. 20,000 + 1,000 + 30 + 7
4. Four hundreds six tens two ones
5. 3 hundred thousands 9 ten thousands 6 hundreds 8 tens 4 ones
6. 7 ten thousands 4 thousands 5 hundreds 7 ones
7. 100,000 + 50,000 + 2,000 + 80
8. 10,000 + 6,000 + 4 + 30 + 300

9. 900 + 700,000 + 60,000 + 40 + 8

10. 7 thousands 6 tens 9 hundred thousands 9 ones 4 hundred

Find the place value of the underlined digit.

11. $7{,}12\underline{8}{,}254$

12. $52{,}918{,}51\underline{2}$

13. $1\underline{2}3{,}403{,}890$

14. $1{,}2\underline{9}5$

15. $2\underline{0}0{,}200$

16. $\underline{9}7{,}290$

Select the correct answer to the following multiple-choice items on place value.

17. Which is read “sixty-five thousand and twenty-one”?
   A. 65, 210
   B. 6,521
   C. 65,021
   D. 650,021

18. Which digit goes in the box to make the statement below true?

$$847{,}286 > 8\square 5{,}300$$

   A. 9
   B. 8
   C. 6
   D. 4

19. The table below shows sales of four different magazines. The number to the right of each magazine is the total number of copies sold. Which magazine sold the most copies?

Magazine A	2,000,120
Magazine B	2,010,210
Magazine C	2,001,100
Magazine D	2,100,100

A. *Magazine A*
B. *Magazine B*
C. *Magazine C*
D. *Magazine D*

20. Tamika read that the population of NJ was 8,724,560. What is the value of the 7 in that number?

A. 70,000
B. 700,000
C. 7,000,000
D. 70,000,000

Answers on pages 169 to 170.

ESTIMATION

Estimation is a vital skill in today's society. Estimation often involves common sense. When estimating, consider what answers could be reasonable and/or round off the numbers in the problem so you can do the arithmetic more easily.

Example 1

If you multiply a 2-digit number by a 2-digit number, you will get a 3- or 4-digit number.

Example 2

$43 \times 98 = ?$ Since 43 is close to 40 and 98 is close to 100; thus a good estimate of the answer is $40 \times 100 = 4{,}000$.

ESTIMATION EXERCISES

Complete the following multiple-choice problems on Estimation.

1. The sum of 31,999 and 45,510 is best described as
   A. Between 10,000 and 20,000
   B. Between 30,000 and 50,000
   C. Between 70,000 and 80,000
   D. Between 100,000 and 200,000

2. Which of the following is the best estimate of the solution to 39×101?
   A. 400
   B. 4,000
   C. 10,000
   D. 400,000

3. The table below shows the number of phone calls a radio station received each week for the past three days. What is the best estimate for the total number of calls received?

Day	Calls Received
1	5,900
2	7,890
3	4,020

   A. 5,000
   B. 10,000
   C. 15,000
   D. 20,000

4. The difference of 41,908 and 29,340 is best described as

   A. A little more than 100

   B. A little more than 1,000

   C. A little more than 10,000

   D. A little more than 100,000

Answers on page 170.

NUMERICAL OPERATIONS

The NJASK5 will require you to

- Recall basic multiplication and division number facts through 12×12.
- Multiply two-digit numbers.
- Use an efficient and accurate pencil-and-paper procedure for division of a three-digit number by a two-digit number.
- Note that fractions and decimals are also required and are discussed in the next section.
- Construct and use procedures for performing decimal addition and subtraction.
- Recognize the appropriate use of addition and subtraction involving fractions and/or mixed numbers in problem situations.
- Construct, use, and explain procedures for performing addition and subtraction with fractions and decimals.

NUMERICAL OPERATIONS EXERCISES

Try the following examples.

1. $4 \times 9 \times 25 =$

2. $900 - 201 =$

3. $54 \times 39 =$

4. $327 \div 3 =$

5. A gallon contains 128 ounces. Gueimero wants to divide two gallons of soda equally among his eight friends. How much soda will each friend receive?

6. The total cost for Angel to attend 6 days of summer camp was $780. The cost for each day of the camp is the same. What is the cost of each day?

7. Amanda is buying hot dogs for a party. She needs 156 hot dogs. The hot dogs come packaged in packs of six. How many packages should she buy?

Answers on page 170.

RATIONAL NUMBERS (FRACTIONS AND DECIMALS)

Fractions and decimals (rational numbers) permeate the entire NJASK5. Thus it is critically important that you have conceptual understanding of rational numbers. A rational number can be written as a fraction $\frac{a}{b}$ such that a and b are integers {. . . , −3, −2, −1, 0, 1, 2, 3, . . .} and $b \neq 0$. The top number or numerator, a, tells how many fractional pieces there are. The bottom number or denominator, b, of a fraction tells how many pieces an object was divided into.

Example 1

$\frac{4}{5}$, 4 is the numerator and 5 is the denominator.

Example 2

Which is larger: $\frac{3}{8}$ or $\frac{5}{8}$? If the denominators of two fractions are the same, the fraction with the largest numerator is the larger fraction. Thus $\frac{5}{8}$ is larger than

$\frac{3}{8}$ because all of the pieces are the same and five pieces are more than three pieces.

Example 3

If two fractions have different numerators and denominators it is difficult to determine which fraction is larger.

$$\text{Which is bigger? } \frac{2}{3} \text{ or } \frac{3}{4}$$

$$\frac{2}{3} \cdot \frac{4}{4} = \frac{8}{12} \text{ and } \frac{3}{4} \cdot \frac{3}{3} = \frac{9}{12}$$

$$\text{Since } \frac{9}{12} > \frac{8}{12}, \text{ then } \frac{3}{4} > \frac{2}{3}$$

Note you can and should use calculators (except for the SCR problems) to compare fractions.

$$\frac{2}{3} = 0.666\ldots \text{ and } \frac{3}{4} = 0.75$$

$$\text{Since } 0.75 > 0.66\ldots, \text{ then } \frac{3}{4} > \frac{2}{3}$$

Decimals, like fractions, are a representation of a rational number. Rational numbers in decimal form will terminate (end) or repeat (like $\frac{1}{3} = 0.33333\ldots$). The decimal 0.5 represents the fraction $\frac{5}{10}$. The decimal 0.25 represents the fraction $\frac{25}{100}$. Decimals always have a denominator based on a power of 10. We know that $\frac{5}{10}$ is equivalent to $\frac{1}{2}$ since $\frac{1}{2} \cdot \frac{5}{5} = \frac{5}{10}$. Similarly, the decimal 0.5 is equivalent to $\frac{1}{2}$ or $\frac{2}{4}$, etc. Some common equivalent decimals and fractions are shown below.

Fraction	0	$\frac{1}{10}$	$\frac{1}{5}$	$\frac{1}{4}$	$\frac{1}{3}$	$\frac{2}{5}$	$\frac{1}{2}$	$\frac{3}{5}$	$\frac{2}{3}$	$\frac{3}{4}$	$\frac{4}{5}$	1
Decimal	0	0.1	0.2	0.25	$0.\bar{3}$	0.4	0.5	0.6	$0.\bar{6}$	0.75	0.8	1

The place value system extends to the right of the decimal point. The number 1234.567 is displayed below.

Thousands	Hundreds	Tens	Ones	Decimal Point	Tenths	Hundredths	Thousandths
1	2	3	4	•	5	6	7

RATIONAL NUMBERS EXERCISES

1. If these fractions were graphed on the number line, which of them would be closest to zero?

   A. $\frac{1}{2}$

   B. $\frac{1}{3}$

   C. $\frac{1}{4}$

   D. $\frac{1}{5}$

2. Which of the following is equivalent to $\frac{3}{4}$?

A. 0.66

B. 0.7

C. 0.75

D. 0.8

3. Which of the following is greater than $\frac{4}{5}$?

A. $\frac{3}{4}$

B. $\frac{5}{6}$

C. $\frac{1}{2}$

D. $\frac{3}{5}$

4. Asafa Powell of Jamaica holds the record for the fastest 100-meter dash. His time was nine and seventy-four hundredths seconds. How is this time written as a number?

A. 974

B. 97.4

C. 9.74

D. 0.974

5. Mrs. Gallia's car holds 12 gallons of gasoline. Mr. Gallia's car holds only 9.5 gallons. How much more does one car hold than the other?

   A. 2.5 gallons

   B. 2 gallons

   C. 1.5 gallons

   D. 0.5 gallon

6. The square below shows a fraction of a whole that is shaded.

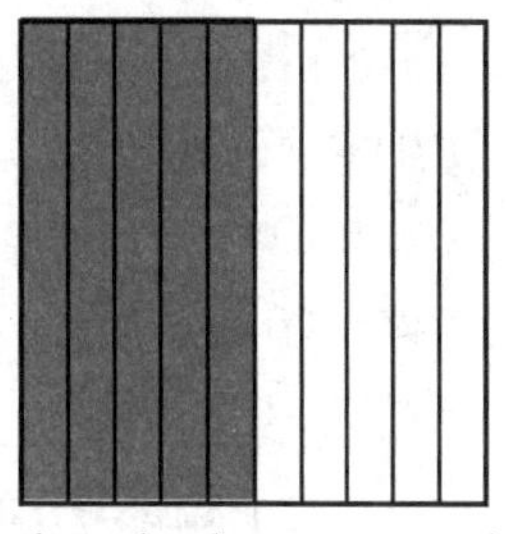

How many of the circles below would have to be shaded to be equivalent to the fraction that is shaded in the square above?

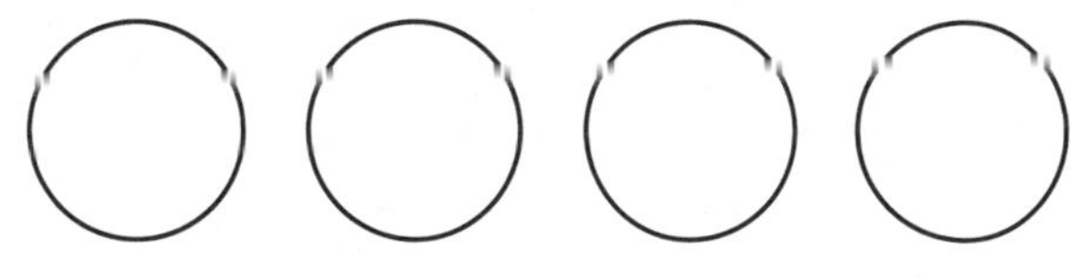

7. If a circle is fully shaded, then it is equal to one whole; find the sum of the shaded area of the two circles below.

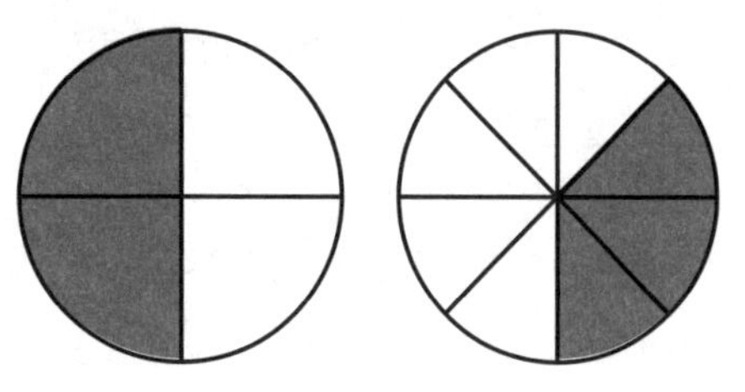

8. Austin has a \$5 bill. He wants to purchase two pretzels for 75¢ each and one soda for 50¢. How much change will Austin receive after purchasing the pretzel and the soda?

   A. \$4.00

   B. \$3.75

   C. \$3.50

   D. \$3.00

9. Ms. Krulik's students rode their bikes at a state park. The sign shows the distance from the entry to the park to each destination. The students rode their bikes to the baseball field and then rode back. What was the total distance that they traveled?

Destination	Distance
Lake	$2\frac{1}{2}$ miles
Picnic area	$3\frac{1}{4}$ miles
Baseball field	$1\frac{3}{4}$ miles

   A. $2\frac{1}{2}$ miles

   B. $3\frac{1}{4}$ miles

   C. $3\frac{1}{2}$ miles

   D. $7\frac{1}{2}$ miles

10. Which of the following shows the numbers listed from least to greatest?

A. $0.5, \frac{3}{5}, 1, 1.01$

B. $0.9, 1\frac{1}{3}, 1, 1.5$

C. $0.7, \frac{3}{5}, 1\frac{1}{2}, 2$

D. $0.5, \frac{1}{5}, 1, 1\frac{1}{2}$

11. At the annual school race, Kaleem ran 3.1 miles and Lisa ran only 1.95 miles in the same amount of time. How much farther did Kaleem run than Lisa?

A. 1 mile

B. 1.1 miles

C. 1.15 miles

D. 1.5 miles

12. What is 0.01 more than 2.05?

A. 2.06

B. 3.05

C. 3.06

D. 3.15

13. A fifth-grade class voted for their favorite subject. Two-thirds of the class of 36 voted that math was their favorite. The rest of the students in the class voted for language arts. How many more students would have to choose mathematics for $\frac{3}{4}$ of the students to like it?

14. Four students have three gallons of soda to split equally among themselves. How much should each of them receive?

15. State a number that is between 0.2 and $\frac{1}{4}$.

Answers on pages 171 to 172.

NUMBER THEORY

Number theory terms such as ***prime***, ***composite***, ***factors***, and ***multiples*** will be on the NJASK5. Several important definitions follow below.

A **prime** number is a whole number greater than one that is evenly divisible by only 1 and itself. In other words, a prime number has exactly two factors.

2, 3, 5, 7, 11, 13, 17, 19, 23, . . .

A **composite** number is a whole number greater than one that is evenly divisible by more than two numbers.

4, 6, 8, 9, 10, 12, 14, 15, 16, . . .

Note that the number 1 is neither prime nor composite. In determining if a given number is prime or composite, you need to decide if a number besides 1 and itself divides evenly into the given number. Divisibility rules can help determine this. Some rules are listed below.

- Divisibility by 2: If a number ends in 0, 2, 4, 6, or 8, it is an even number.
- Divisibility by 3: If the sum of the digits of a number is divisible by 3, then the number is divisible by 3.
- Divisibility by 5: If a number ends in 5 or 0, it is divisible by 5.
- Divisibility by 10: If a number ends in 0, it is divisible by 10.

Factoring is the process of writing a number as the product of other numbers.

$$12 = \{1, 2, 3, 4, 6, 12\}$$

A factor tree is a diagram that can be used to factor whole composite numbers into the product of primes.

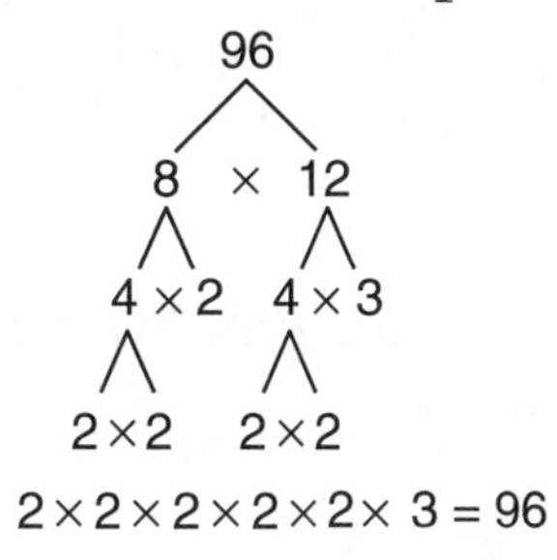

NUMBER THEORY EXERCISES

1. How many numbers between 7 and 41 have no remainder when divided by 5?

   A. 7 numbers

   B. 8 numbers

   C. 9 numbers

   D. 10 numbers

2. Which of the answers below shows the prime factorization of 90?

   A. 2×45

   B. $2 \times 5 \times 9$

   C. $2 \times 3 \times 5$

   D. $2 \times 3 \times 3 \times 5$

3. Which is a prime factor of the number 36?

   A. 2

   B. 4

   C. 6

   D. 9

4. Ms. Stewart had a class trip. There was 1 adult for every 8 students. If a total of 72 students went on the trip, how many adults were needed?

   A. 1 adult

   B. 8 adults

   C. 9 adults

   D. 72 adults

5. A number that has exactly two factors is called _____.

   A. composite

   B. prime

   C. factor

   D. multiple

6. Which list contains no prime numbers?

   A. 2, 4, 6, 8

   B. 4, 8, 12, 16

   C. 5, 9, 18, 20

   D. 10, 15, 19, 20

7. The least common multiple of 4 and 6 is _____?

   A. 2

   B. 8

   C. 12

   D. 24

EXTENDED CONSTRUCTED-RESPONSE ITEMS

1. On a number line, plot points for the following numbers.

$$\frac{3}{5} \text{ and } 0.8$$

- Label each point.
- Name two different rational numbers that are greater than $\frac{3}{5}$ and less than 0.8. (Write one of your numbers in fractional form and write the other number in decimal form.)
- Explain how you know that each of your numbers is greater than $\frac{3}{5}$ and less than 0.8.

2. The Magell Company makes many different sizes of balls. The balls get many different stickers placed on them. Richard places a triangle on every third ball. Justin places a square on every fourth ball, and Melissa places a hexagon on every sixth ball. If 100 balls were made yesterday, then:
   - How many balls had a triangle?
   - How many balls had both a triangle and a square?
   - Would the 24th ball that was made yesterday have any shapes painted on it? If so, how many and what were they?
   - Explain how you found your answers.

3. Lisa thinks that 57 and 67 are both prime because they both end in 7. Is she correct? Explain why or why not?

Chapter 4

GEOMETRY AND MEASUREMENT

Spatial sense is an intuitive feel for shape and space. Geometry and measurement both involve describing the shapes we see all around us in art and nature and the things we make. Spatial sense, geometric modeling, and measurement can help us to describe and interpret our physical environment and to solve problems.

In grade 5, test items come from five areas as follows:

Geometric Properties. This includes

- Identifying, describing, and classifying standard geometric objects
- Describing and comparing properties of geometric objects
- Explaining symmetry, congruence, and similarity

Transforming Shapes. This includes

- Combining shapes to form new ones
- Decomposing complex shapes into simpler ones
- Using standard geometric transformations of translation (slide), reflection (flip), rotation (turn), and dilation (scaling)

Coordinate Geometry. Coordinate geometry provides an important connection between geometry and algebra.

- In grade 5, students need only to know the first quadrant of the coordinate plane.

Units of Measurement. Measurement helps describe our world using numbers. This includes

- Attaching numbers to real-world phenomena
- Demonstrating familiarity with common measurement units (e.g., inches, liters, and miles per hour)

Measuring Geometric Objects. This area focuses on applying the knowledge and understandings of units of measurement in order to actually perform measurement.

- In grade 5, students need to be familiar with the perimeter and area of squares and rectangles only.

You should realize that geometry and measurement are all around us. Through study of these areas and their applications, you should come to better understand and appreciate the role of mathematics in your life.

TYPES OF LINES

A **line** is a straight path that goes on forever in both directions.

A **line segment** is a section of a line between two points called its endpoints.

A **ray** is a straight path that begins at an endpoint and goes on forever in one direction.

TYPES OF ANGLES

An angle is formed by two rays that share a common endpoint. This shared endpoint is called the **vertex** of the angle. An angle is named by a point on one ray, the vertex point, and a point on the other ray. An angle can also be named by the vertex point. The notation for angle is ∠.

Example

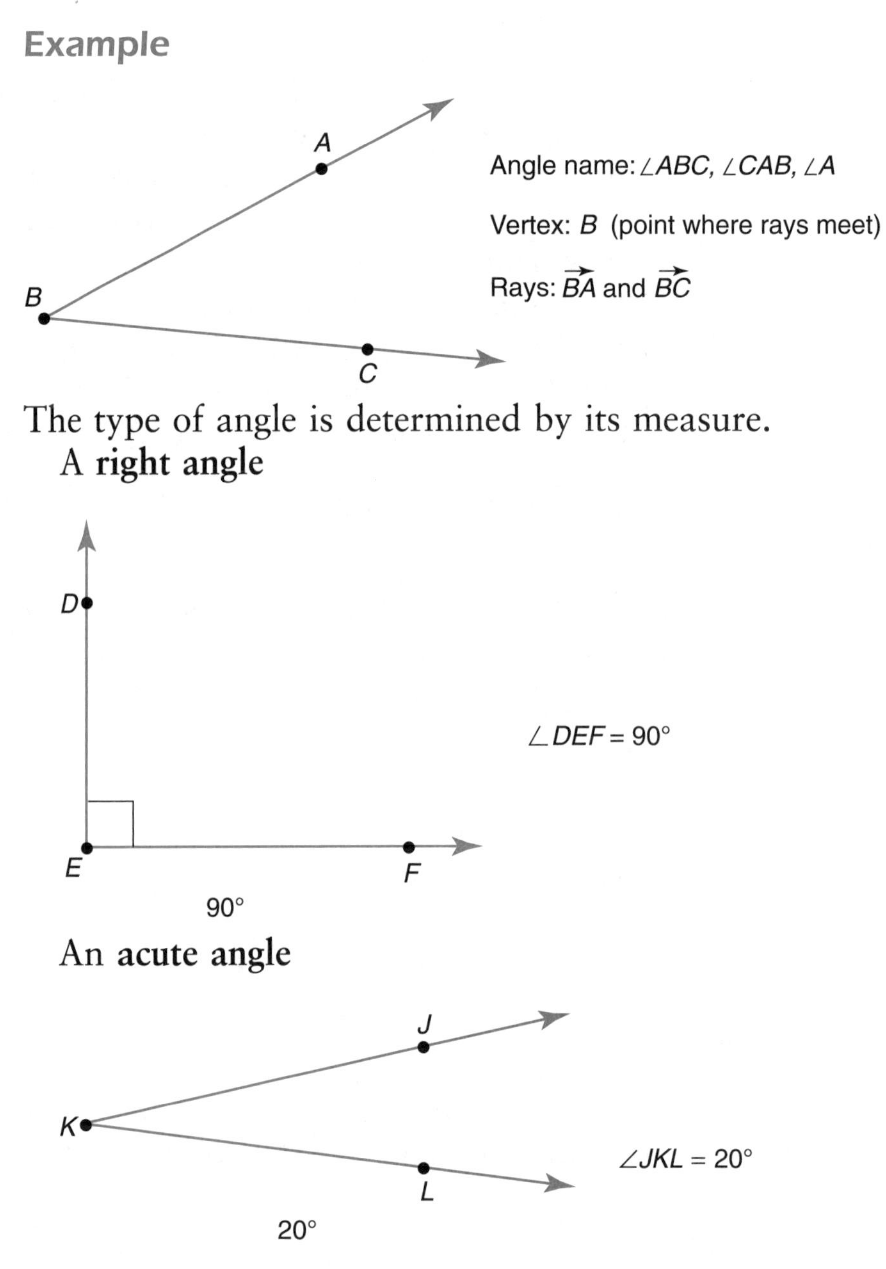

The type of angle is determined by its measure.

A **right angle**

An **acute angle**

An **obtuse angle** measures between 90° and 180°.

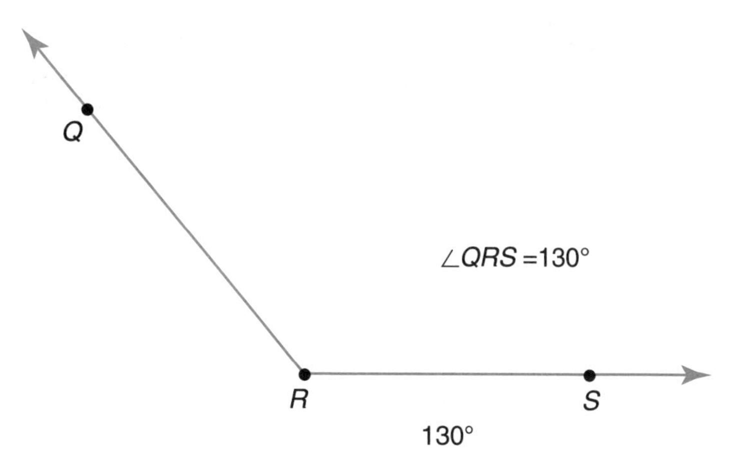

TYPES OF ANGLES EXERCISES

Use the following angle to answer the questions below.

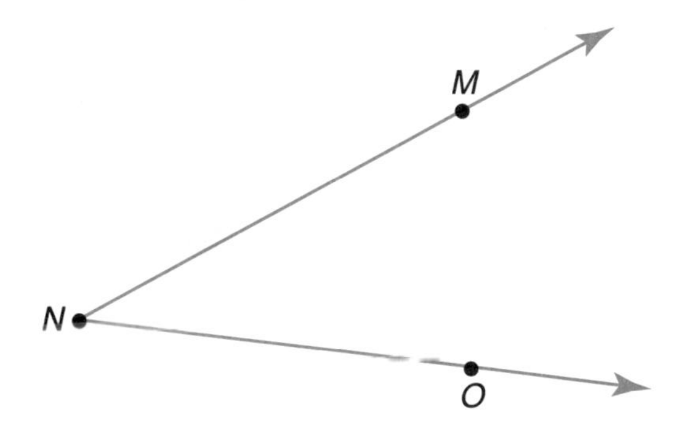

1. Name the above angle in two ways using the correct symbols.

2. What is the vertex?

3. List the rays.

4. Identify the type of angle.

Answers on page 173.

PARALLEL LINES

Parallel lines are lines on a flat surface that never intersect and are always the same distance apart. An example is shown below:

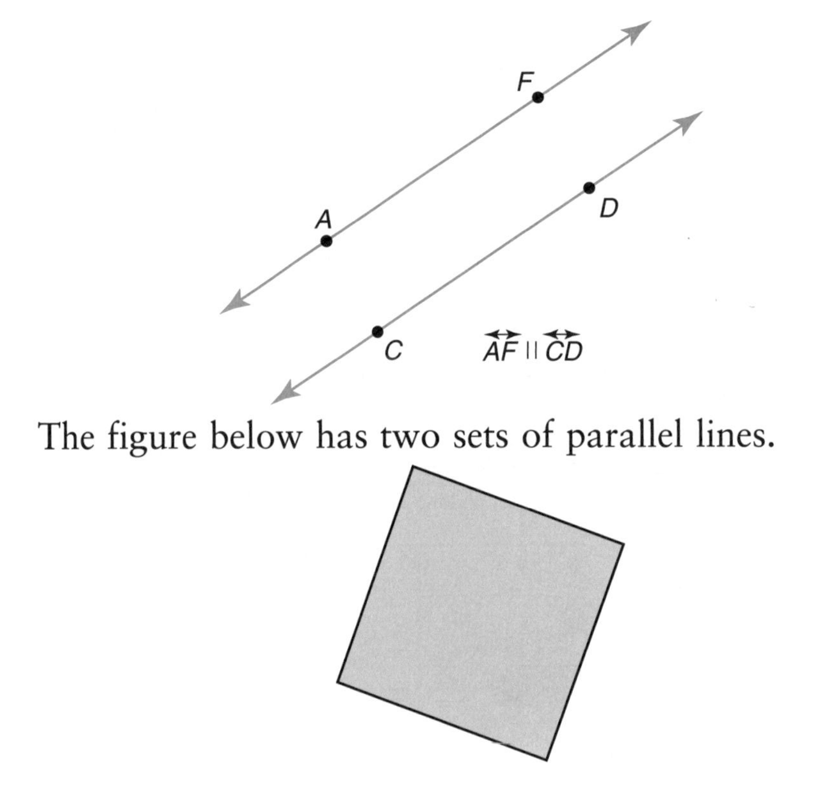

The figure below has two sets of parallel lines.

PARALLEL LINES EXERCISES

How many sets of parallel lines are in each figure below?

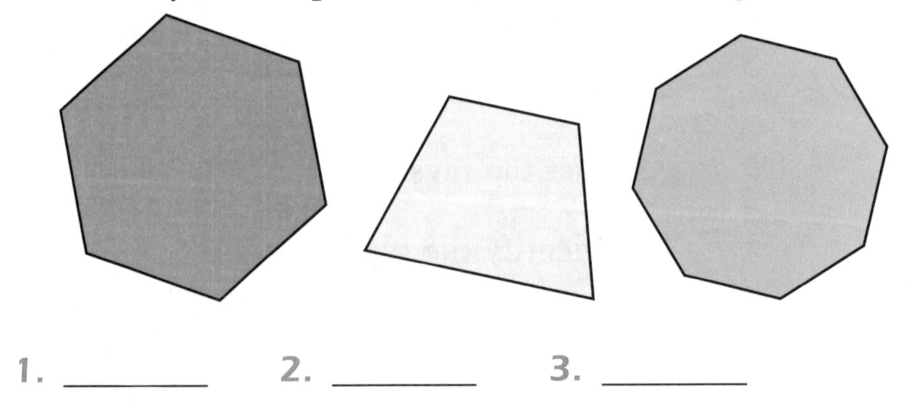

1. ________ 2. ________ 3. ________

Answers on page 173.

PERPENDICULAR LINES

Perpendicular lines are lines that intersect to form right angles. See the figure below.

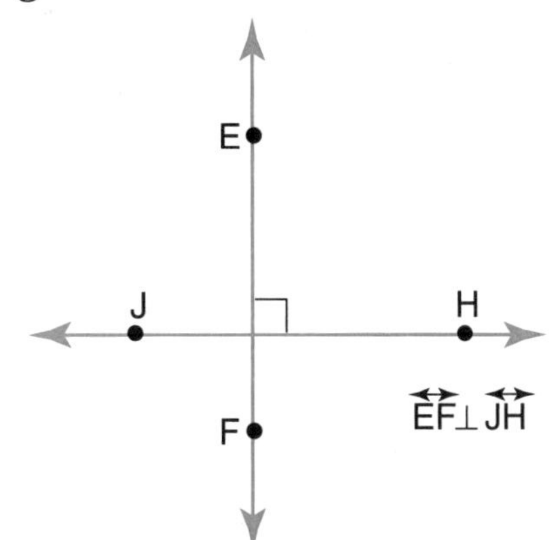

Example. The figure below has four sets of perpendicular lines.

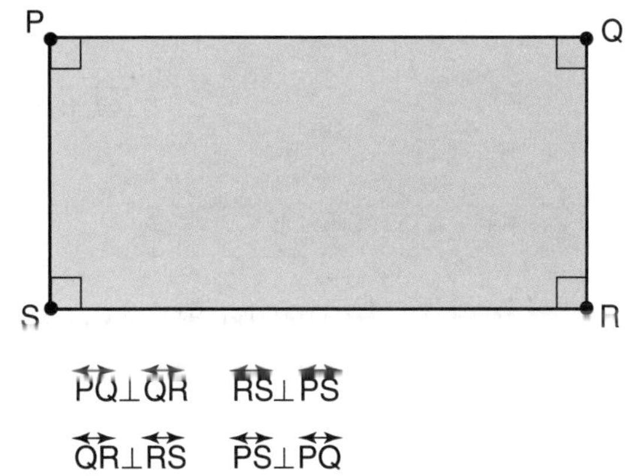

PERPENDICULAR LINES EXERCISES

How many sets of perpendicular lines are in each figure below?

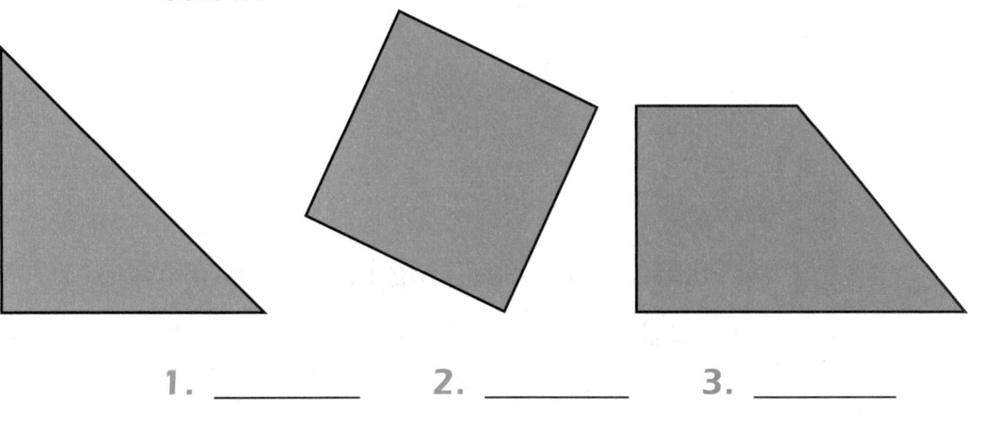

1. ________ 2. ________ 3. ________

Answers on page 173.

INTERSECTING LINES

Intersecting lines are lines that cross each other at exactly one point.

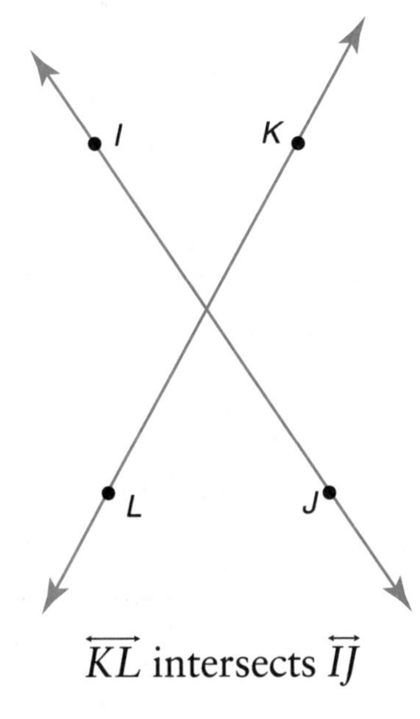

$\overleftrightarrow{KL}$ intersects $\overleftrightarrow{IJ}$

INTERSECTING LINES EXERCISES

Use this figure to answer qucstions 1–3.

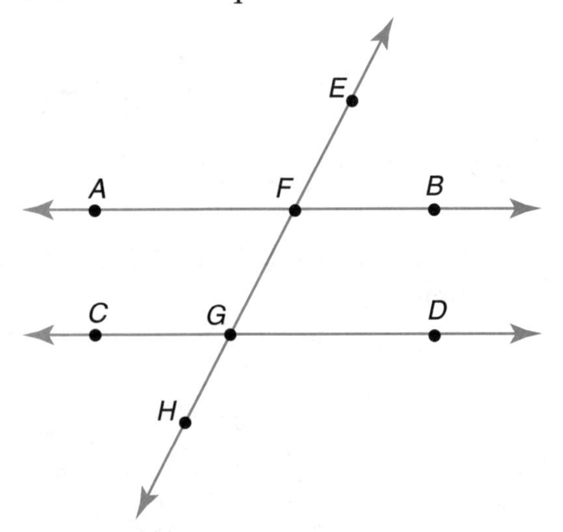

1. Which of the following lines do not intersect?

   A. $\overleftrightarrow{AB}$ and $\overleftrightarrow{EF}$

   B. $\overleftrightarrow{CD}$ and $\overleftrightarrow{EH}$

   C. $\overleftrightarrow{AB}$ and $\overleftrightarrow{CD}$

2. Name a pair of parallel lines.

3. Name three pairs of intersecting lines.

4. Using the figure above name four pairs of perpendicular lines.

Answers on pages 173 to 174.

POLYGONS

A **polygon** is a closed, two-dimensional figure with at least three sides. The sides of a polygon are made up of line segments that meet at endpoints called vertices. These vertices never intersect.

TRIANGLES

A **triangle** is the simplest polygon because it has three sides. Triangles can be classified by side lengths.

Scalene Triangle	Isosceles Triangle	Equilateral Triangle
No equal sides	At least two sides are equal	All sides are equal

Triangles can also be classified by angle measures.

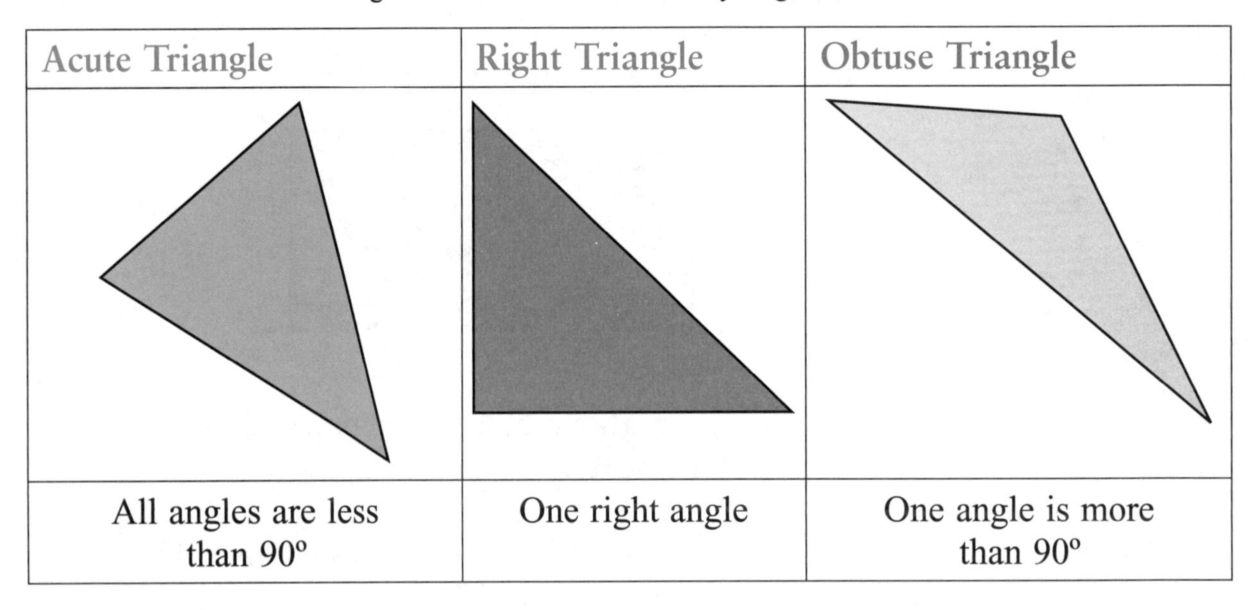

Acute Triangle	Right Triangle	Obtuse Triangle
All angles are less than 90°	One right angle	One angle is more than 90°

TRIANGLES EXERCISES

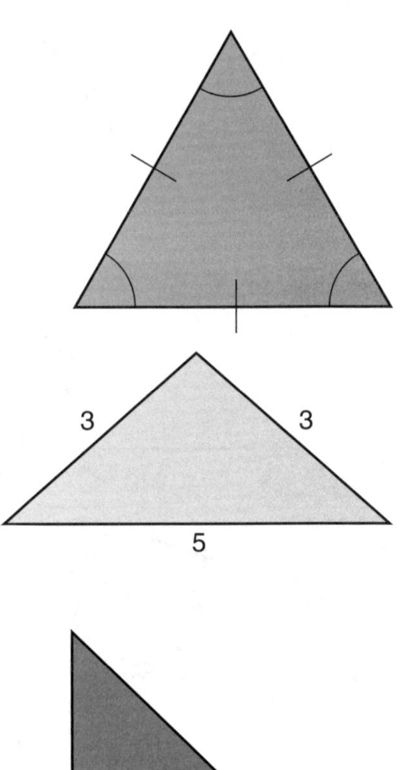

1. What kind of triangle is this?
   - A. Acute triangle
   - B. Right triangle
   - C. Equilateral triangle
   - D. Obtuse triangle

2. What kind of triangle is this?
   - A. Right triangle
   - B. Isosceles triangle
   - C. Equilateral triangle
   - D. Obtuse triangle

3. What kind of triangle is this?
   - A. Right triangle
   - B. Isosceles triangle
   - C. Equilateral triangle
   - D. Obtuse triangle

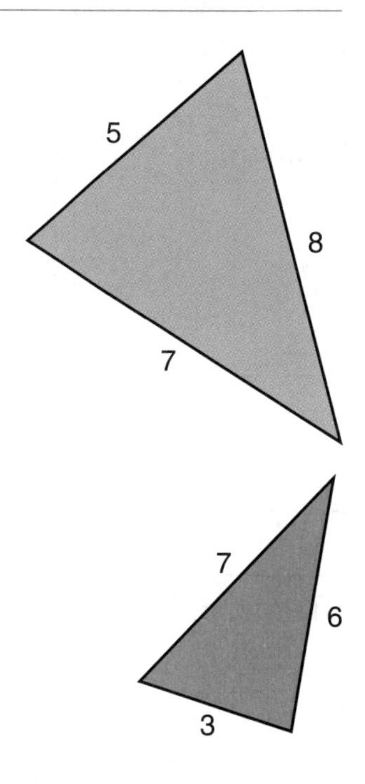

4. What kind of triangle is this?
   A. Right triangle
   B. Scalene triangle
   C. Isosceles triangle
   D. Equilateral triangle

5. What kind of triangle is this?
   A. Right triangle
   B. Acute triangle
   C. Obtuse triangle
   D. Scalene triangle

Answers on page 174.

QUADRILATERALS

A **quadrilateral** is a polygon with four sides. Another name for quadrilateral is **quadrangle**. All quadrangles have four angles, four vertices, and four sides.

A quadrilateral with exactly two pairs of parallel sides is a **parallelogram**. These quadrilaterals are parallelograms.

Name	Figure	Description
Parallelogram		Opposite sides have the same length.
Rectangle		Opposite sides have the same length and all four angles are right angles.
Rhombus		All four sides have the same length.

Name	Figure	Description
Square		All four sides have the same length, and all four angles are right angles. All squares are rectangles and rhombuses.

These quadrilaterals are NOT parallelograms.

Name	Figure	Description
Trapezoid		Has exactly one pair of parallel sides. All four sides of a trapezoid can have different lengths.
Isosceles trapezoid		Has exactly one pair of parallel sides and one pair of equal sides.

QUADRILATERALS EXERCISES

1. A rhombus is _____.
   A. equiangular
   B. equilateral
   C. obtuse
   D. regular

2. Which of the following is not a parallelogram?
   A. Trapezoid
   B. Square
   C. Rectangle
   D. Rhombus

3. A square is also a(n) _____.

   A. trapezoid

   B. rhombus

   C. rectangle

   D. isosceles trapezoid

4. What kind of trapezoid is this?

   A. Acute trapezoid

   B. Obtuse trapezoid

   C. Isosceles trapezoid

   D. Equilateral trapezoid

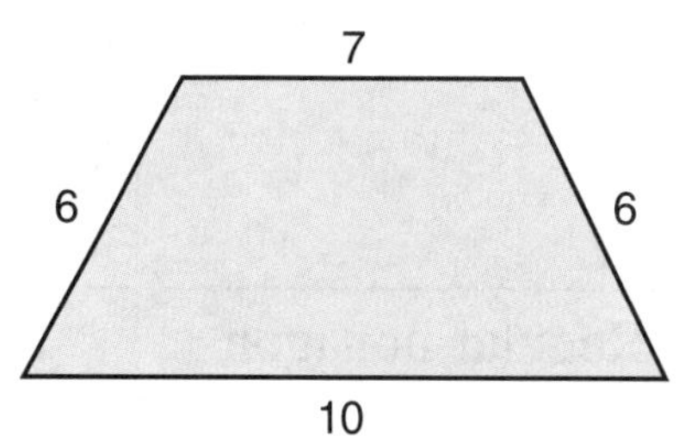

Answers on page 174.

REGULAR POLYGONS

A polygon is **regular** if all of its side lengths are the same and all of its angles are equal in measure. Therefore, regular polygons are equilateral, having equal sides and equiangular, having equal angles. The following are regular polygons.

Name	Figure	Description
Equilateral triangle		Three equal sides and angles.
Square		Four equal sides and angles.
Regular pentagon		Five equal sides and angles.

Name	Figure	Description
Regular hexagon		Six equal sides and angles.
Regular octagon		Eight equal sides and angles.
Regular nonagon		Nine equal sides and angles.

REGULAR POLYGONS EXERCISES

1. A polygon with equal side lengths is _____.
   A. scalene
   B. isosceles
   C. equilateral
   D. quadrilateral

2. A polygon with angles of equal measure is called _____.
   A. acute
   B. equiangular
   C. obtuse
   D. right

3. A polygon with equal side lengths and angles of equal measure is called _____.

   A. isosceles

   B. acute

   C. obtuse

   D. regular

4. A polygon with five sides is a(n) _____.

   A. nonagon

   B. octagon

   C. hexagon

   D. pentagon

5. A polygon with eight sides is a(n) _____.

   A. nonagon

   B. octagon

   C. hexagon

   D. pentagon

6. Which of the following may not be equiangular?

   A. Square

   B. Pentagon

   C. Rectangle

   D. Equilateral triangle

7. Which of the following is always a regular polygon?

   A. Octagon

   B. Hexagon

   C. Pentagon

   D. Square

8. A polygon with nine sides is a(n) _____.

   A. pentagon

   B. hexagon

   C. octagon

   D. nonagon

9. Which polygon below is not a quadrilateral?

   A. Trapezoid

   B. Rhombus

   C. Pentagon

   D. Rectangle

10. A triangle has angles measuring 30°, 60°, and 90°. What kind of triangle is it?

    A. Right triangle

    B. Acute triangle

    C. Scalene triangle

    D. Isosceles triangle

11. William is plotting a parallelogram on a coordinate grid. He has plotted three coordinates so far and needs help with the fourth vertex. At what coordinate should he plot the fourth vertex?

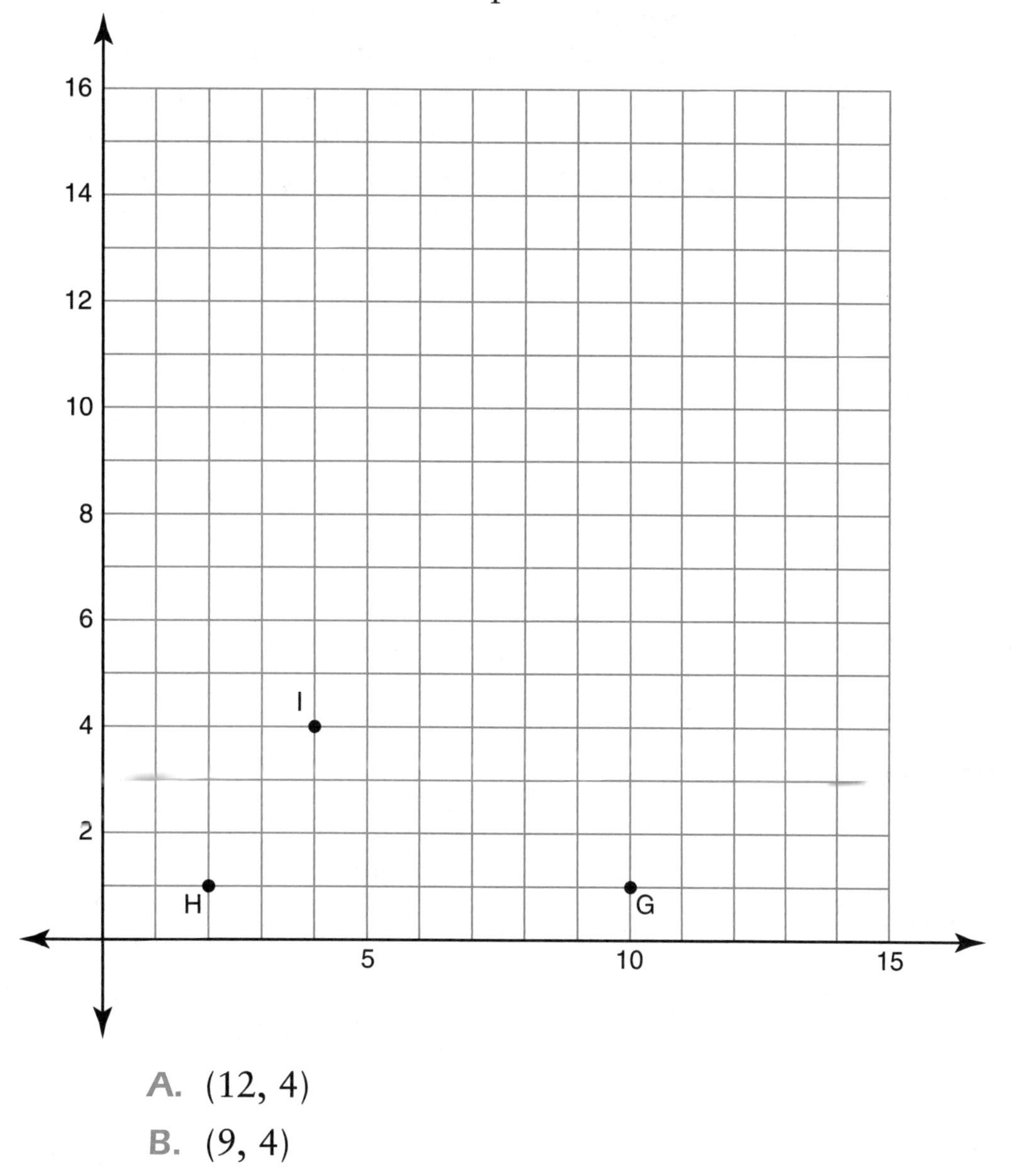

A. (12, 4)
B. (9, 4)
C. (8, 4)
D. (10, 4)

Answers on pages 174 to 175.

CIRCLES

A **circle** is a plane shape enclosed by a set of points that are the same distance away from the center point. The **radius** (*r*) is a line segment that connects the center point of a circle with any point on the circle. The **diameter** (*d*) is a line segment that passes through the center of the circle and connects two points on the circle. The **circumference** (C) of a circle is the distance around the circle.

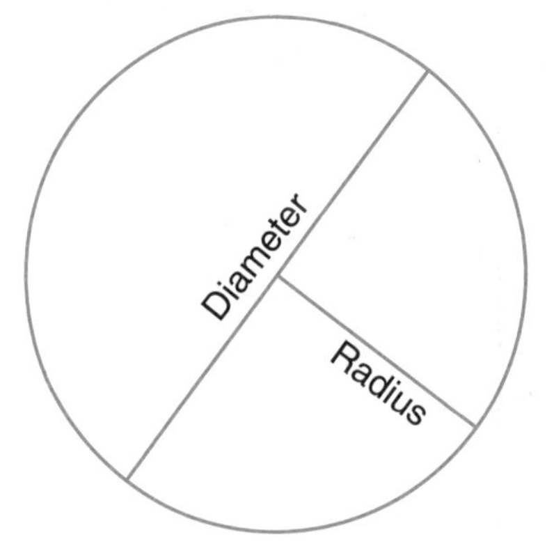

Example 1. The circle below is shown on a coordinate grid.

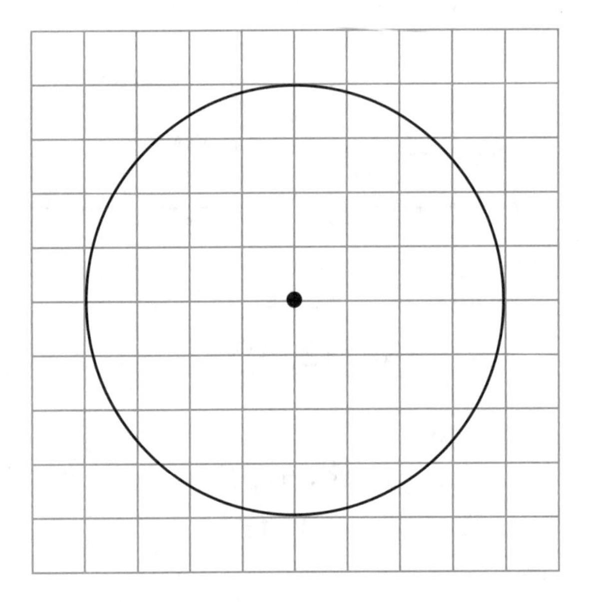

The diameter is 8 units across and the radius $= \left(\frac{1}{2}\right)$ times the diameter = 4 units.

Example 2.

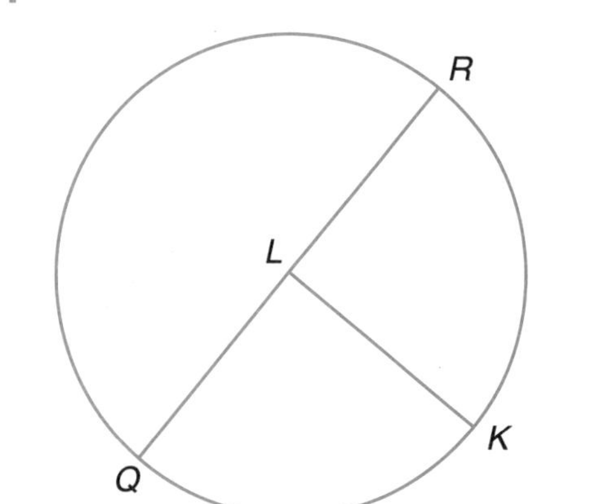

If the radius is 10 units, what is the diameter? The diameter is 20 units. The diameter is two times the radius. The radius is half of the diameter.

CIRCLES EXERCISES

1. Name the diameter of this circle.
2. Name the radius of this circle.
3. Name another radius of this circle.
4. Name another radius of this circle.
5. If the diameter is 8 units, what is the radius?
6. If the radius of another circle is 7 units, what is the diameter?

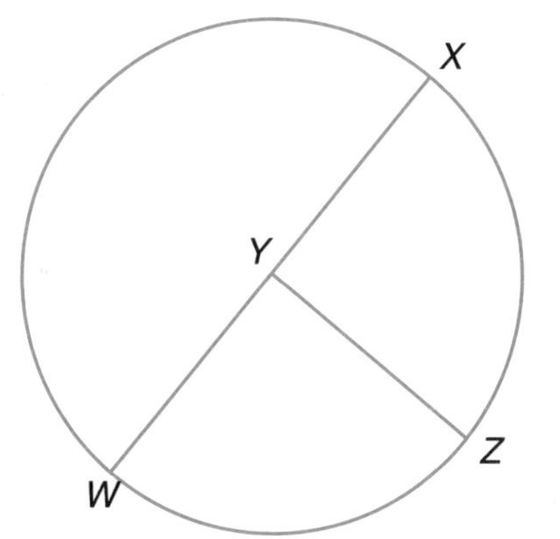

Answers on page 175.

CONGRUENT FIGURES

Two figures that have the same shape and size are **congruent.**

Examples.

Each pair of figures is congruent because the figures in each pair have the same shape and size.

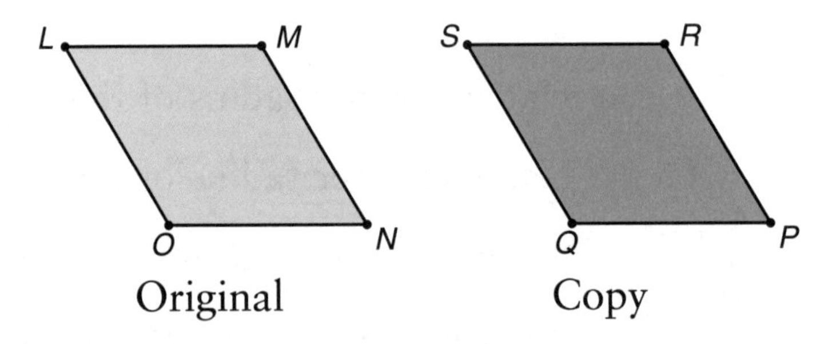

Original Copy

The original figure and the copy are congruent.

CONGRUENT FIGURES EXERCISES

1. Which two triangles below appear to be congruent?

______

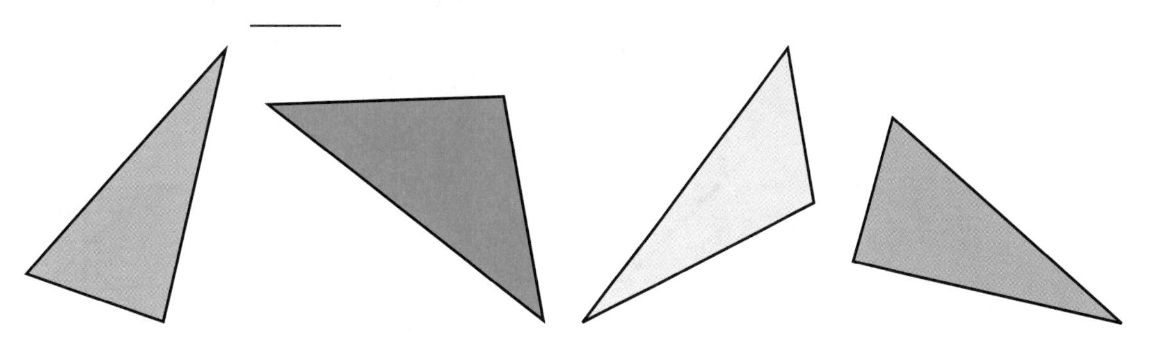

2. Which pair of triangles below is congruent?

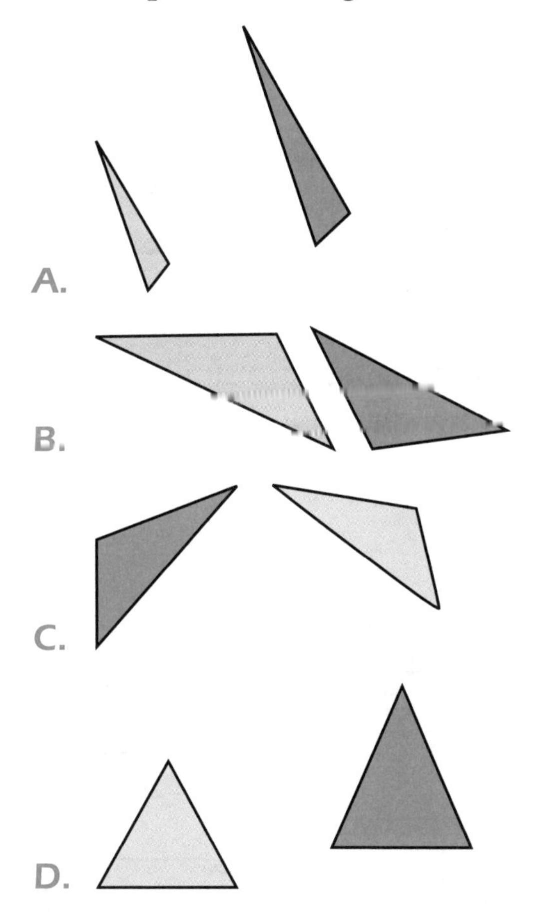

Answers on page 175.

SIMILARITY

Two figures are mathematically **similar** if they have the same shape. Several examples are shown below.

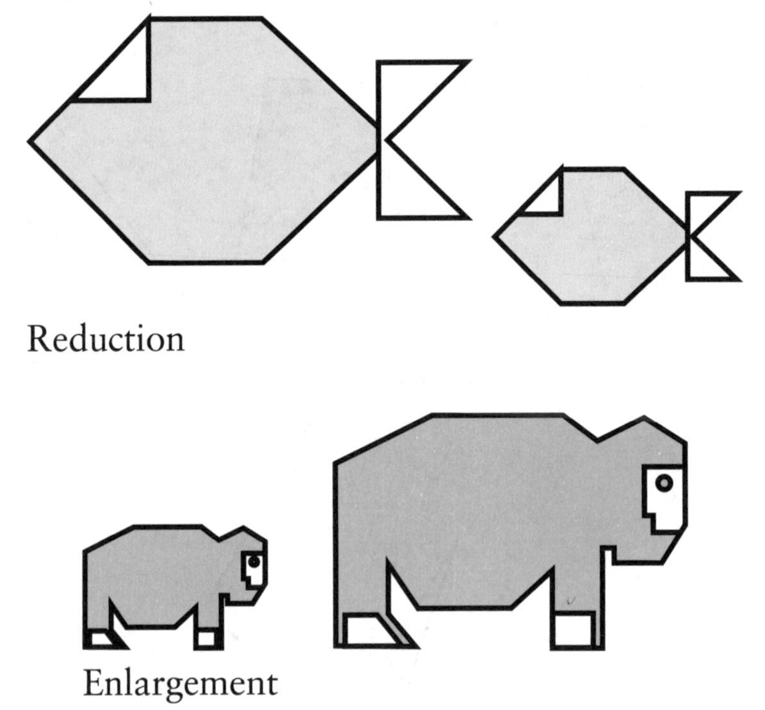

Reduction

Enlargement

Both pairs of figures are similar. **Similarity** means that each pair has the same shape but different sizes.

SIMILARITY EXERCISES

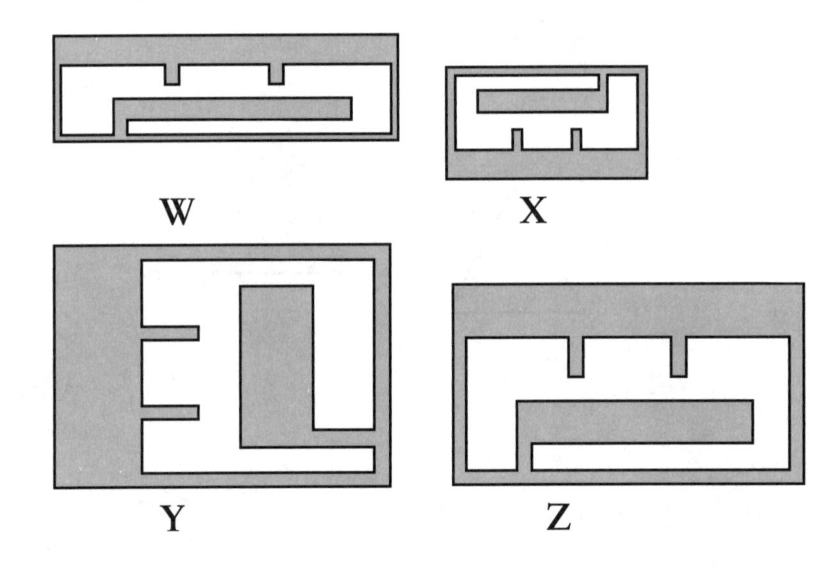

1. Which of the above shapes are similar? _____

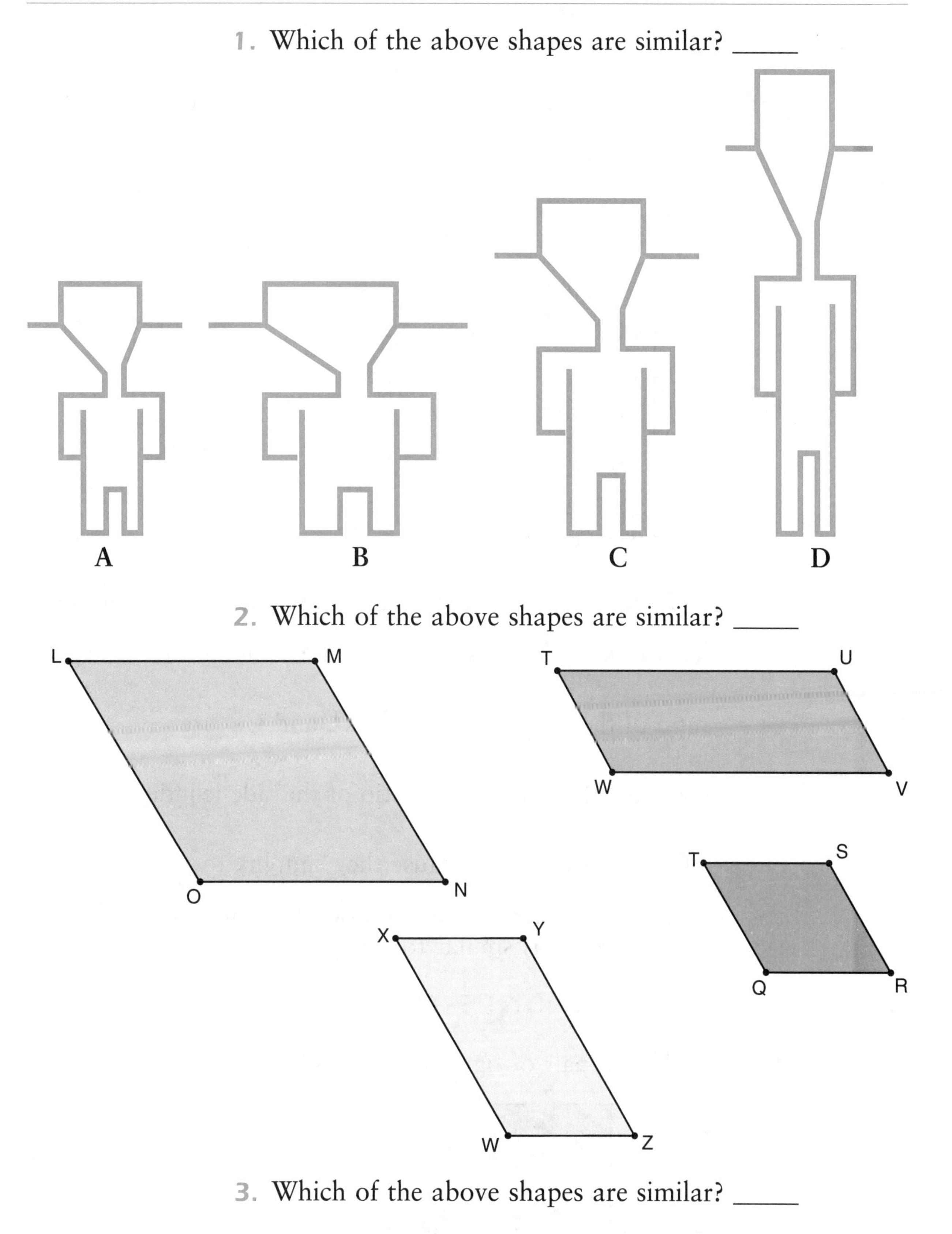

2. Which of the above shapes are similar? _____

3. Which of the above shapes are similar? _____

Answers on page 175.

SIMILARITY AND SIDE LENGTH

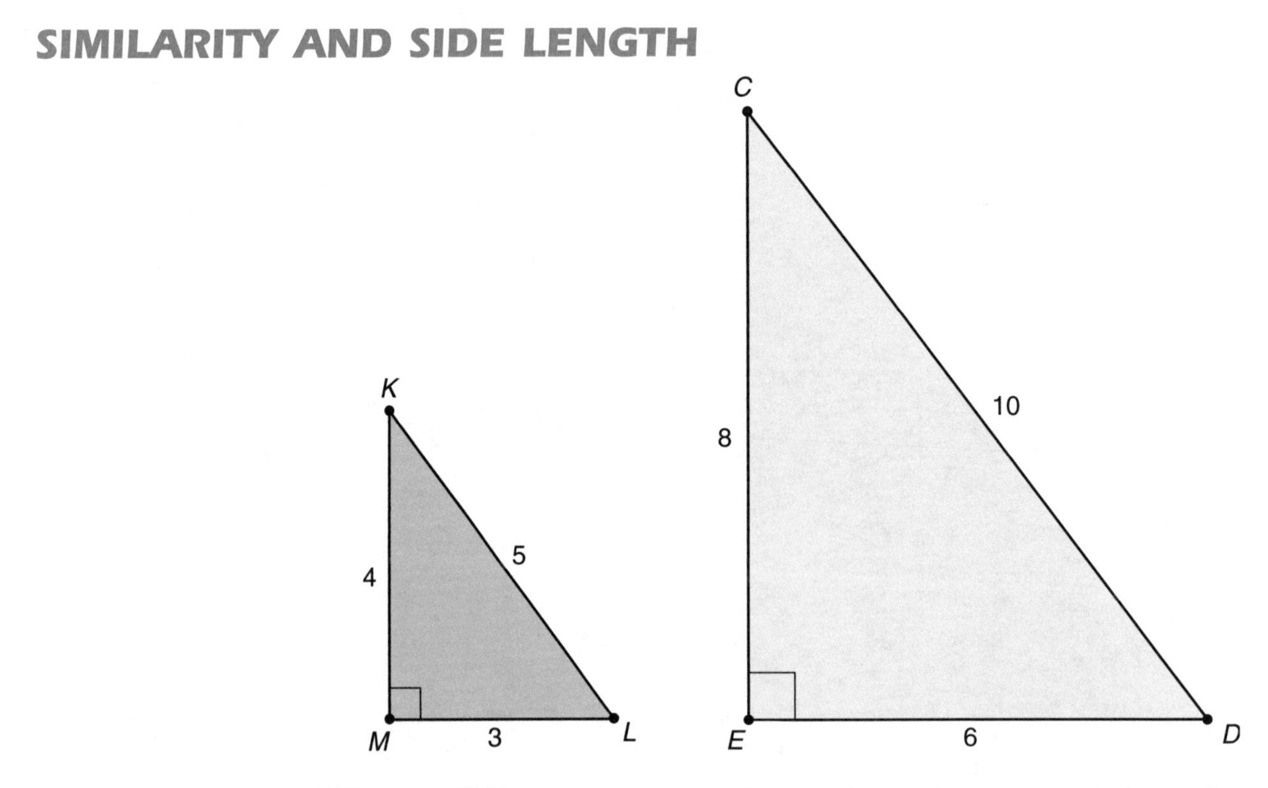

$\overline{LM}$ and $\overline{DE}$ are corresponding sides. The ratio of the side lengths is $\frac{3}{6}$. $\overline{KM}$ and $\overline{CE}$ are corresponding sides. The ratio of the side lengths is $\frac{4}{8}$. $\overline{KL}$ and $\overline{CD}$ are corresponding sides. The ratio of the side lengths is $\frac{5}{10}$. All ratios are equal because they simplify to $\frac{1}{2}$. The lengths of the **corresponding sides** have the same **ratio.** Triangle ***KLM*** is similar to triangle ***CDE.***

SIMILARITY AND SIDE LENGTH EXERCISE

1. Which pair of figures is similar?

A. 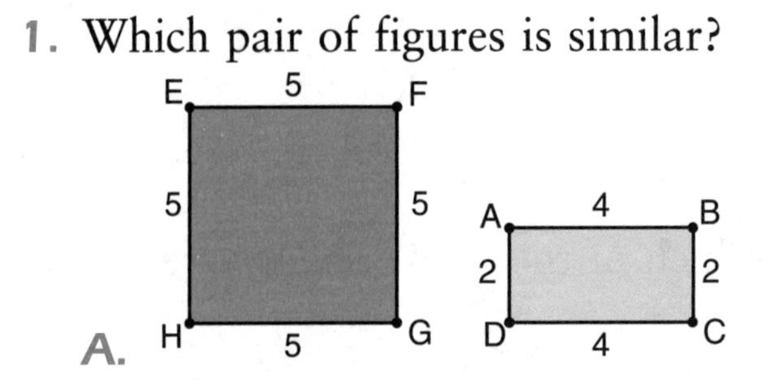

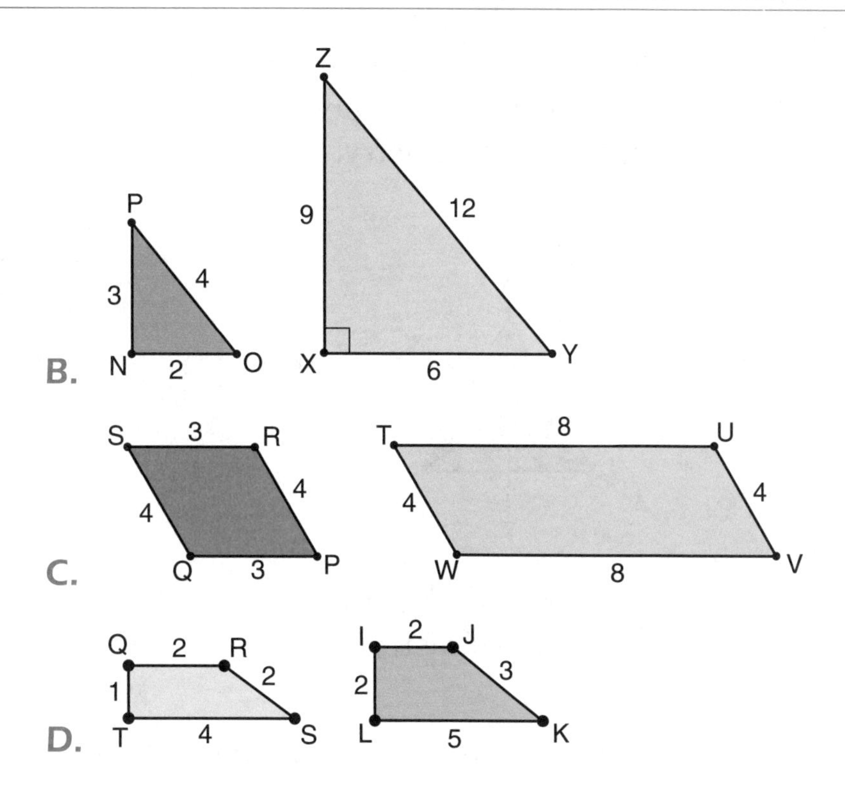

Answers on page 175.

SYMMETRY

An object has line symmetry when it can be divided in halves by one or more lines. Once divided, one half is an exact match of the other half.

Example.

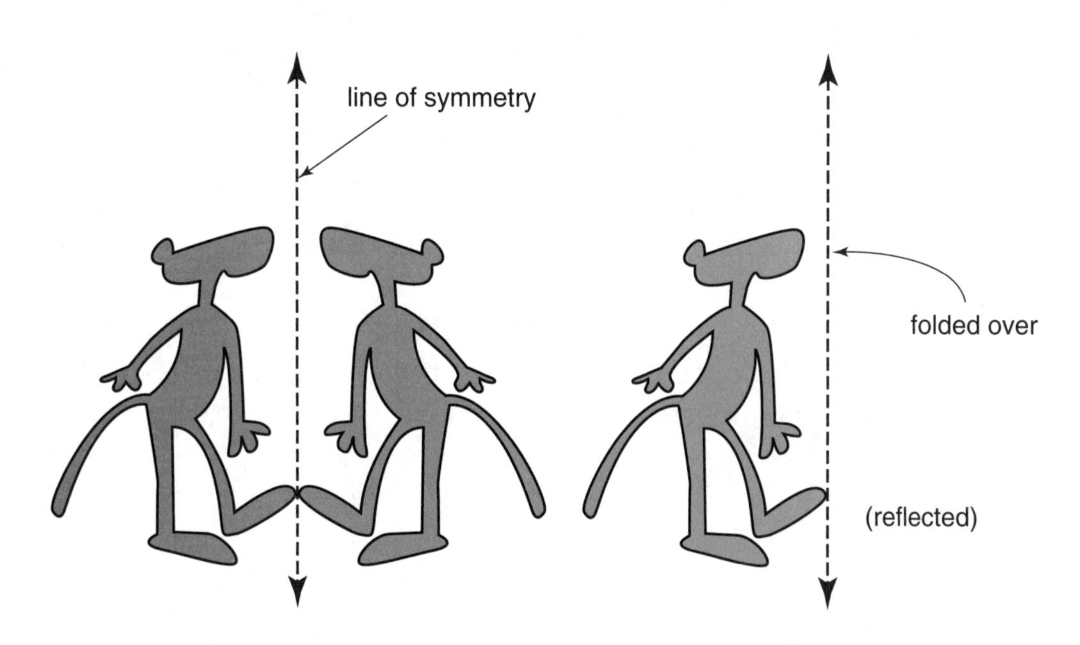

The two figures above can be folded over the line of symmetry so that both halves are an exact match. They have **line symmetry**.

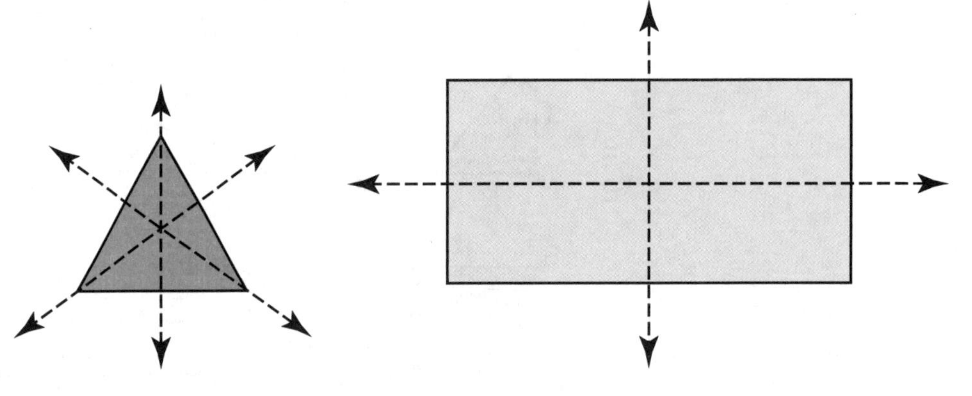

This triangle has three lines of symmetry.

This rectangle has two lines of symmetry.

A figure has **rotation symmetry** if there is a fixed point around which the figure is turned and it still looks the same. The figure must match the original.

Example 1.

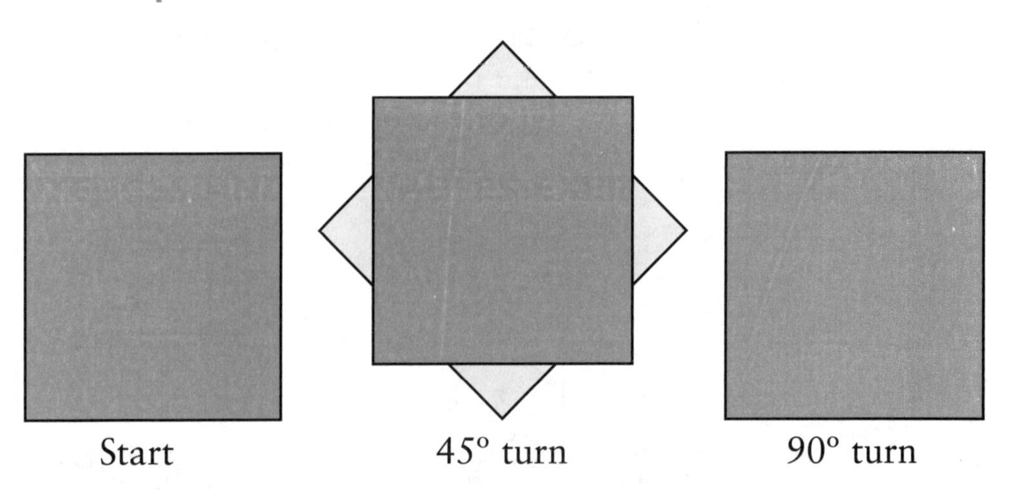

The square matches itself after being rotated 90° around the center point. Therefore the square has rotation symmetry.

Example 2.

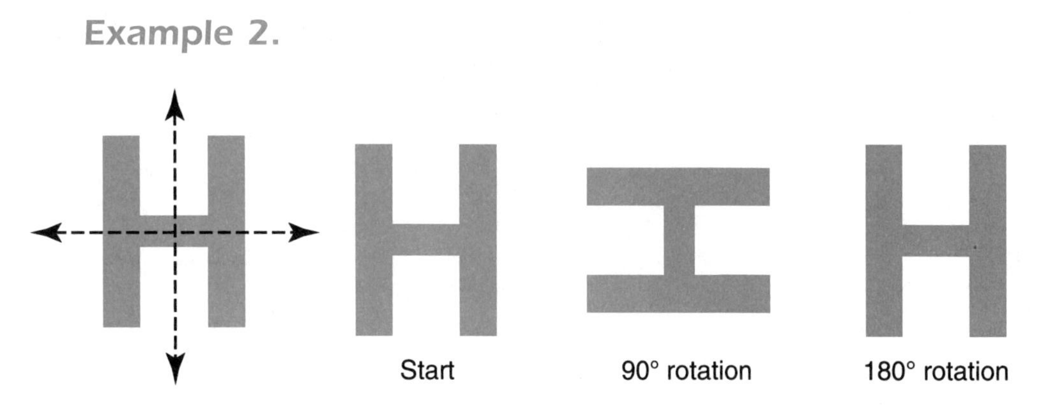

Letter H has both line symmetry and rotation symmetry.

SYMMETRY EXERCISES

1. What kind(s) of symmetry do these letters have? (Letters may have line and rotation symmetry.) Draw in their line(s) of symmetry.

________ ________ ________

2. Use the line of symmetry to construct the matching half of the figure.

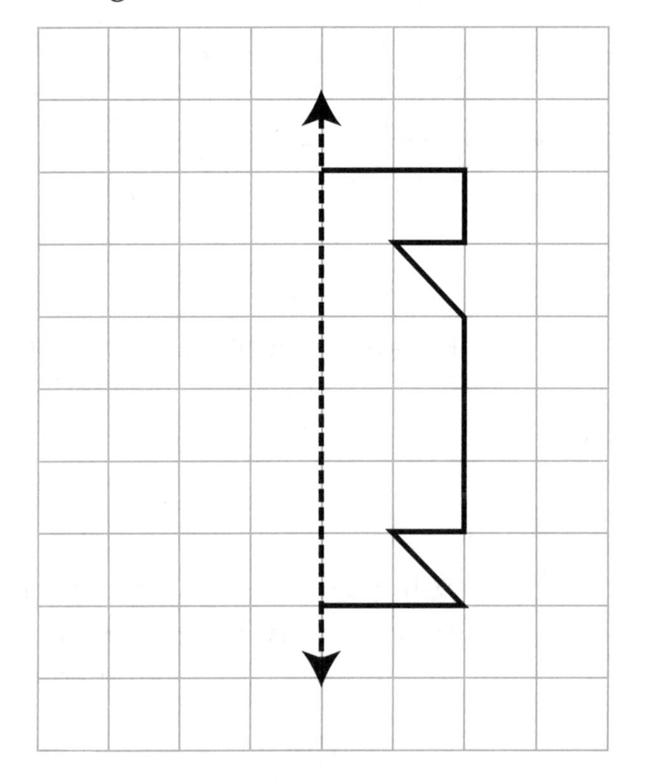

3. Which of the figures below have rotational symmetry?

________ ________ ________

4. Which of the figures below have line symmetry? If so, draw in the line(s) of symmetry.

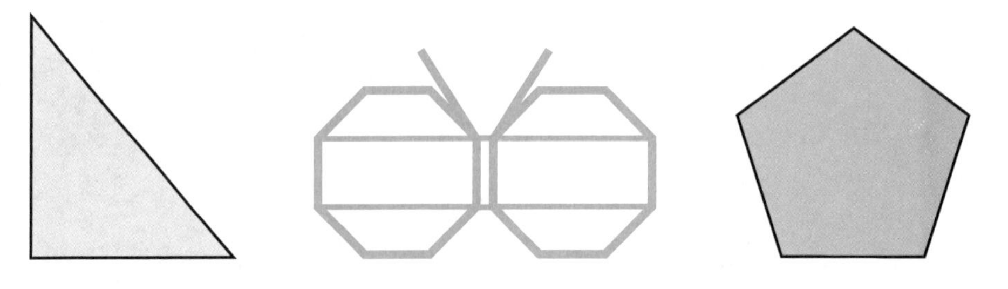

________ ________ ________

Answers on pages 176 to 177.

TRANSFORMATIONS

A figure can be moved from one place to another in three different ways. The new figures are always congruent to the original figure.

A **reflection** is a flip of a figure over a line.

A **translation** is sliding a figure to a new location.

A **rotation** is turning a figure around a point.

Example 1. In a **reflection,** a figure is flipped over a line called the **line of reflection.** Each point on the figure and each corresponding point on the image is the same distance away from the line of reflection. The image is a mirror copy of the original.

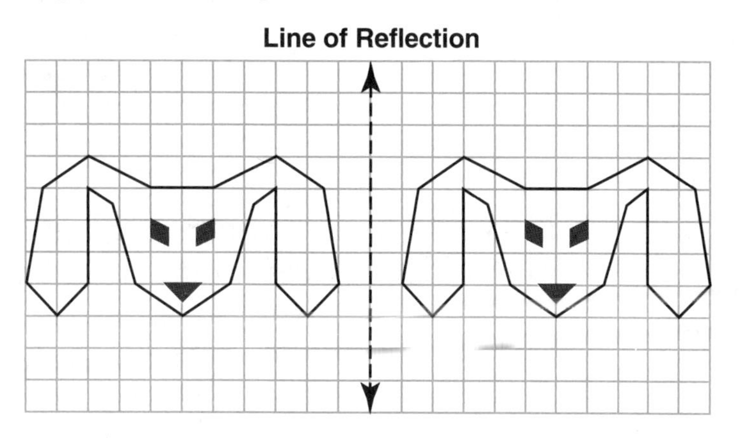

Example 2. In a **translation,** each point on the figure slides the same distance and direction, resulting in the image.

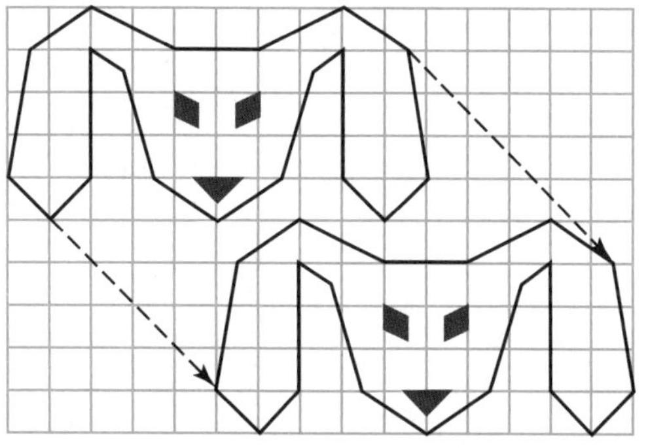

In this example of a **diagonal translation** each point on the original figure slides down 5 units and right 5 units.

In this example of a **horizontal translation** each point on the original figure slides right 12 units.

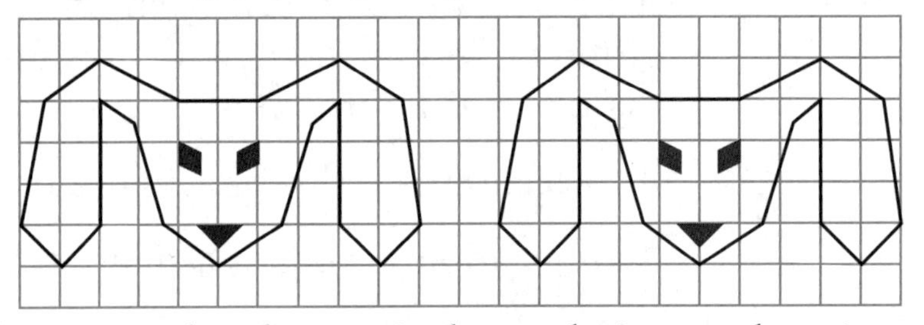

In this example of a **vertical translation** each point on the original figure is translated up 5 units.

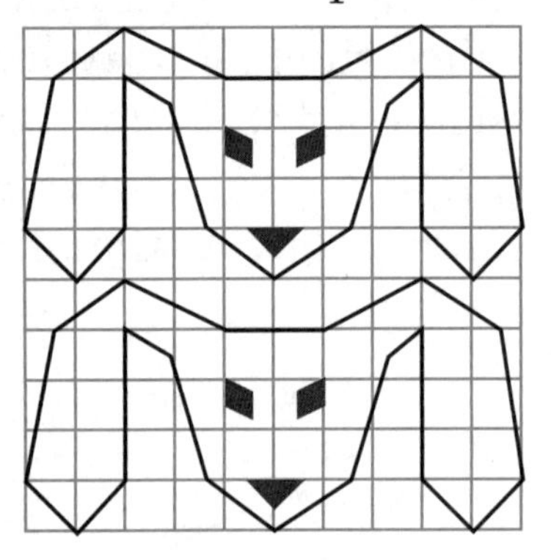

Example 3. In a **rotation,** a figure is turned a certain number of degrees around a point.

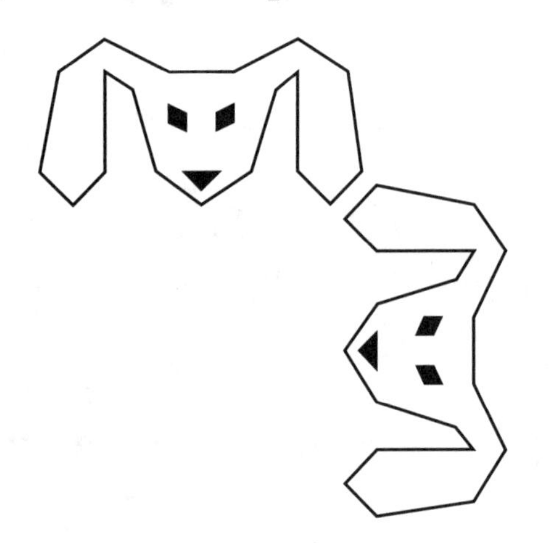

TRANSFORMATIONS EXERCISES

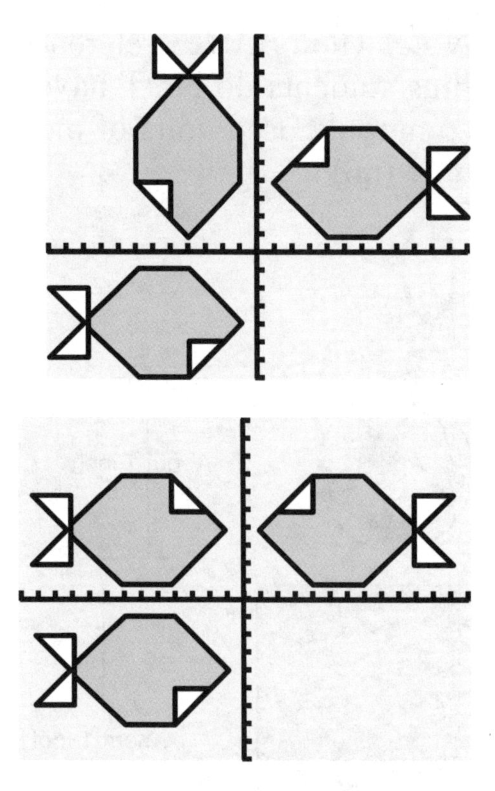

1. What type of transformation is this?
   - A. Reflection
   - B. Rotation
   - C. Translation
   - D. Dilation

2. What type of transformation is this?
   - A. Reflection
   - B. Rotation
   - C. Translation
   - D. Dilation

3. What type of transformation is this?
   - A. Reflection
   - B. Rotation
   - C. Translation
   - D. Dilation

Answers on page 177.

MEASUREMENT

Note: The NJASK5 reference sheet will list **ALL** necessary conversions and thus students do NOT have to memorize conversions. The four most commonly used units of measure are **inch (in.), foot (ft), yard (yd),** and **mile (mi)**

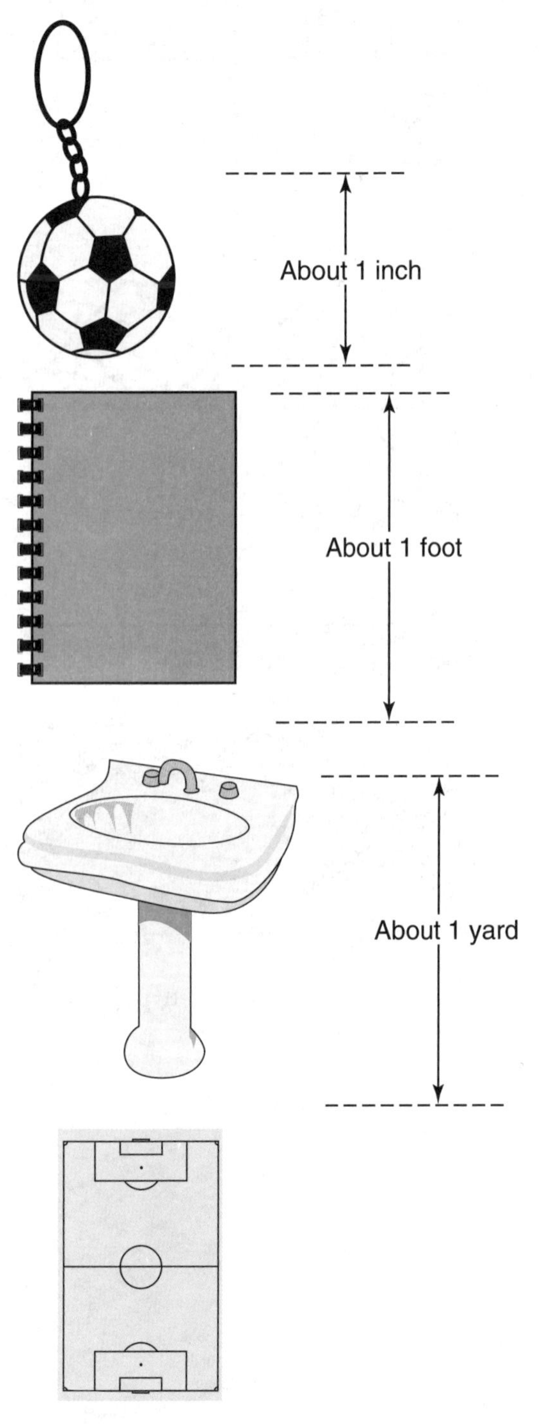

1 inch (in.)

The diameter of the soccer ball on this keychain is about 1 inch.

1 foot (ft)

The length of a spiral notebook is about 1 foot.

1 yard (yd)

The height of a bathroom sink is about 1 yard.

1 mile (mi)

The length of 18 soccer fields is about 1 mile.

18 soccer fields = 1 mile

CONVERTING TRADITIONAL UNITS OF LENGTH

You can convert from one unit to another unit by multiplying or dividing.

To convert from a larger unit to a smaller unit you have to multiply (×).

To convert from a smaller unit to a larger unit you have to divide (÷).

Examples.

Convert	Into	How to	Example
feet	inches	Since there are 12 inches in a foot, multiply by 12	5 ft = 60 in.
yards	feet	Since, there are 3 feet in a yard, multiply by 3	8 yd = 24 ft
miles	yards	Since there are 1,760 yards in a mile, multiply by 1,760	4 mi = 7,040 yd

Do the opposite to convert back. For example, to convert inches into feet divide by 12.

CONVERTING TRADITIONAL UNITS OF LENGTH EXERCISES

1. 9 ft = _____ in.
2. 5,280 yd = _____ mi
3. 84 in. = _____ ft
4. 9 yd = _____ ft
5. 159 ft = _____ yd
6. 12 ft = _____ in.
7. 78 yd = _____ ft
8. 5,280 ft = _____ mi

Answers on page 178.

METRIC UNITS

The four most commonly used metric units of measurement used to measure length are **millimeters (mm)**, **centimeters (cm)**, **meter (m)**, and **kilometer (km).**

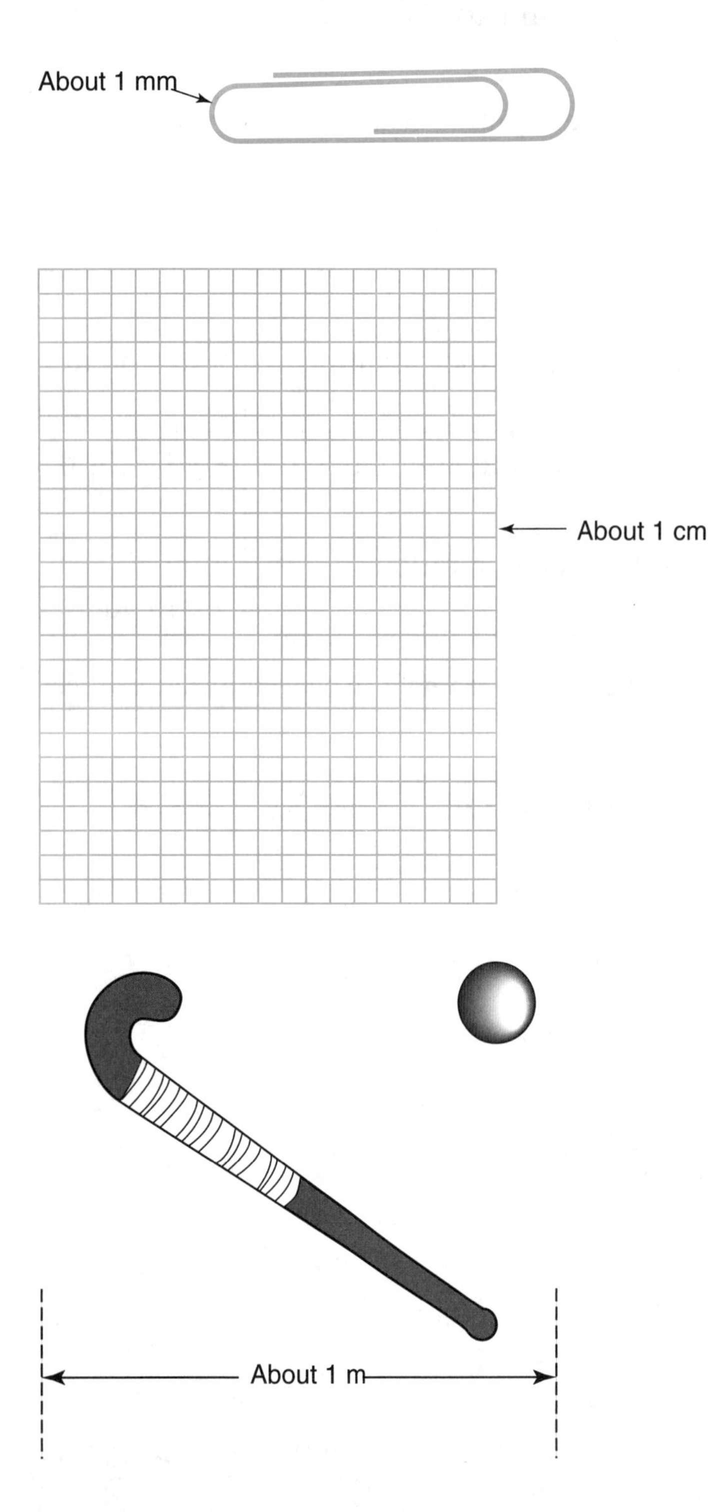

1 millimeter (mm)

The thickness of a paperclip is about 1 millimeter.

1 centimeter (cm) = 10 mm

The squares on graph paper are about 1 centimeter.

1 meter (m) = 100 cm

A field hockey stick is about 1 meter long.

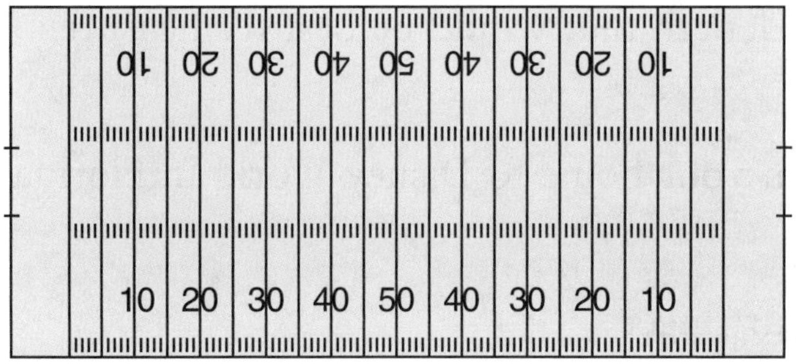

1 kilometer (km) = 1,000 m

The length of 9 football fields placed back to back is about 1 kilometer.

CONVERTING METRIC LENGTH

You can convert from one metric unit to another metric unit by multiplying or dividing

To convert from a smaller unit to a larger unit, you have to divide (÷).

To convert from a larger unit to a smaller unit, you have to multiply (×).

Examples.

Convert	Into	Operation	Example
meters	centimeters	multiply by 100	7 m = 700 cm
centimeter	meters	divide by 100	5 cm = 0.05

CONVERTING METRIC LENGTH EXERCISES

1. 4 m = _____ cm

2. 9 cm = _____ m

3. 1.5 m = _____ cm

4. 545 cm = _____ m

5. 1,000 m = _____ cm

6. 10 cm = _____ m

Directions: Write the metric unit of length that would be best to measure the following.

7. Distance from your home to Disney World in Florida _____

8. Length of your finger _____

9. Height of a basketball pole _____

10. Width of your nail _____

Answers on page 178.

MEASURING LENGTH

One of the most common ways to measure length is to use an inch or centimeter ruler. The inch ruler is divided into 16 parts.

One part represents $\frac{1}{16}$ of an inch.

Two parts are equal to $\frac{2}{16} = \frac{1}{8}$.

Three parts are equal to $\frac{3}{16}$.

Four parts are equal to $\frac{4}{16} = \frac{1}{4}$.

Five parts are equal to $\frac{5}{16}$.

Six parts are equal to $\frac{6}{16} = \frac{3}{8}$.

Seven parts are equal to $\frac{7}{16}$.

Eight parts are equal to $\frac{8}{16} = \frac{1}{2}$.

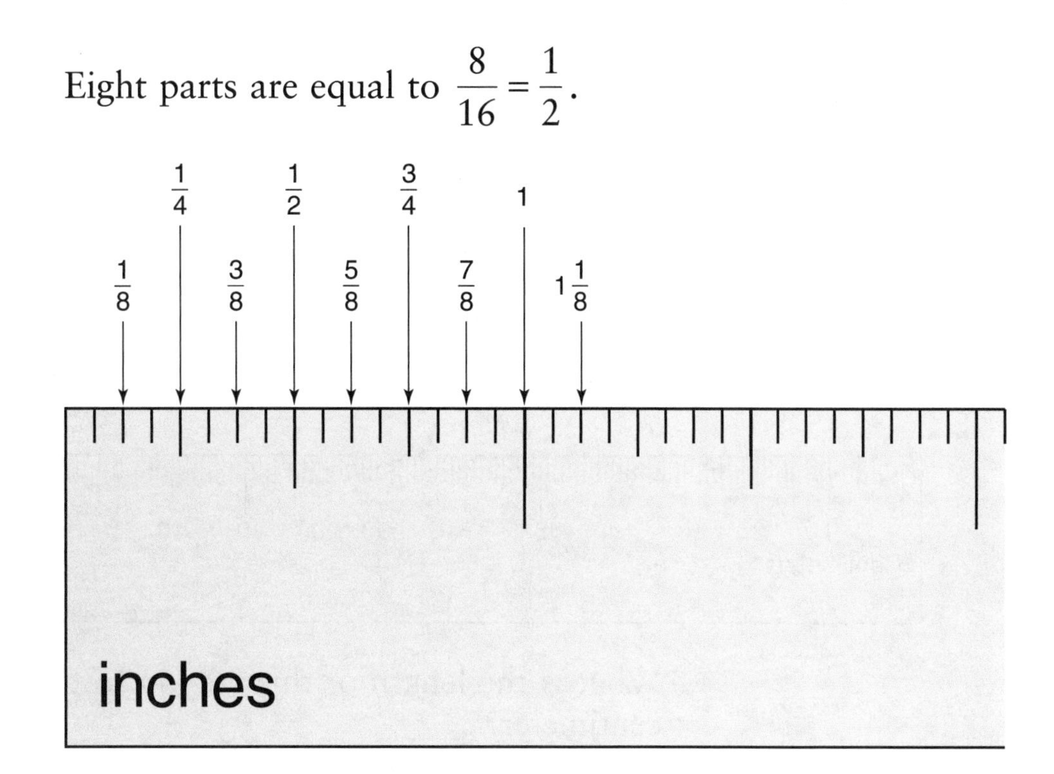

Example.

What is the length of this toy car? (Round to the nearest $\frac{1}{4}$ inch.)

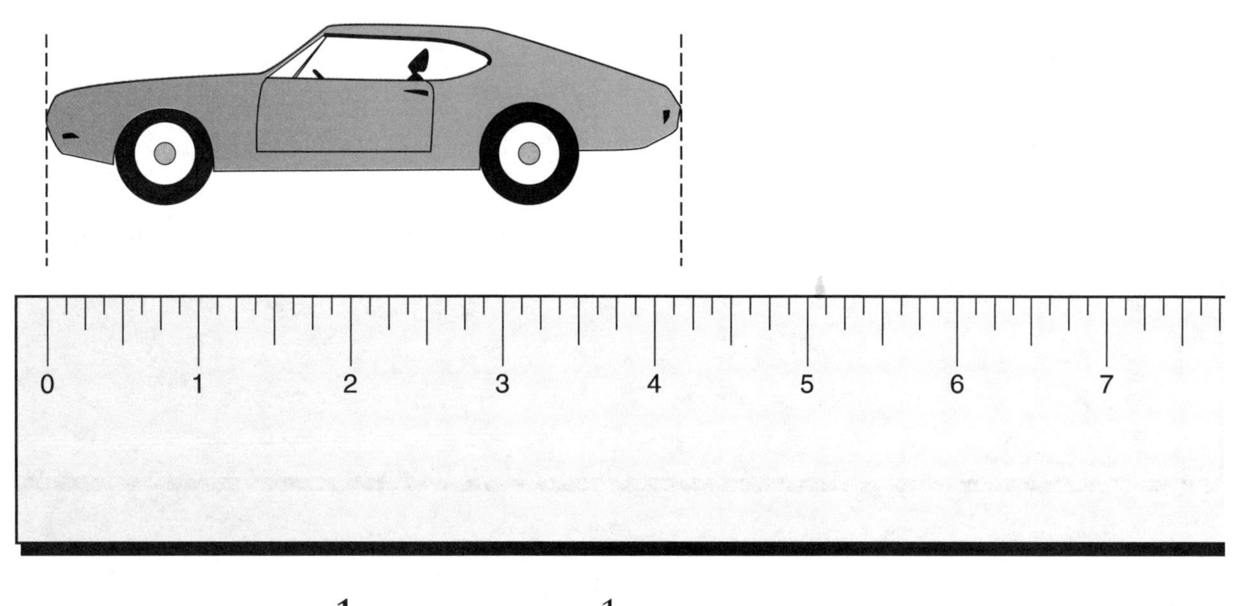

$4\frac{1}{4}$ is the nearest $\frac{1}{4}$ inch.

MEASURING LENGTH EXERCISES

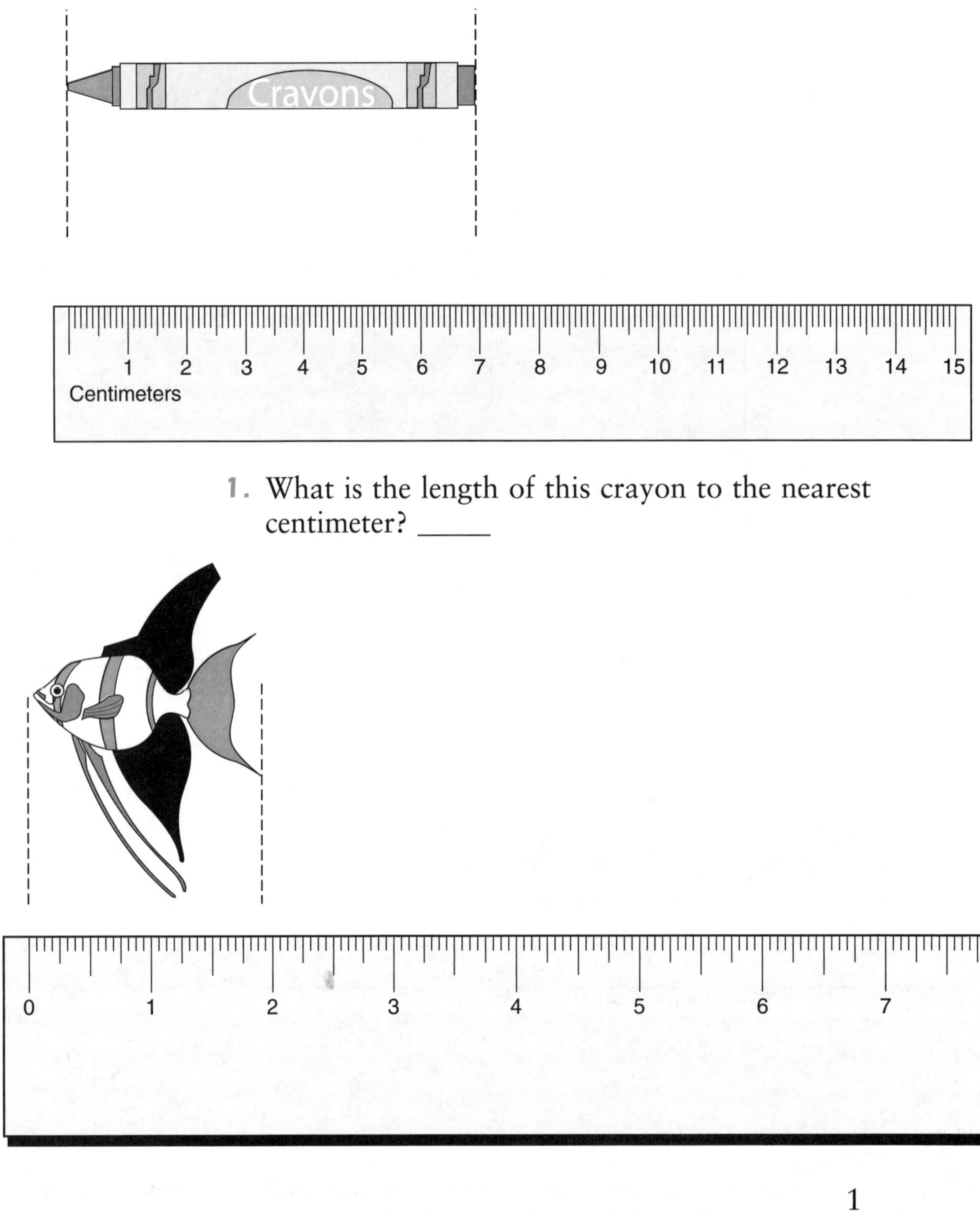

1. What is the length of this crayon to the nearest centimeter? _____

2. What is the length of this fish to the nearest $\frac{1}{4}$ inch? _____

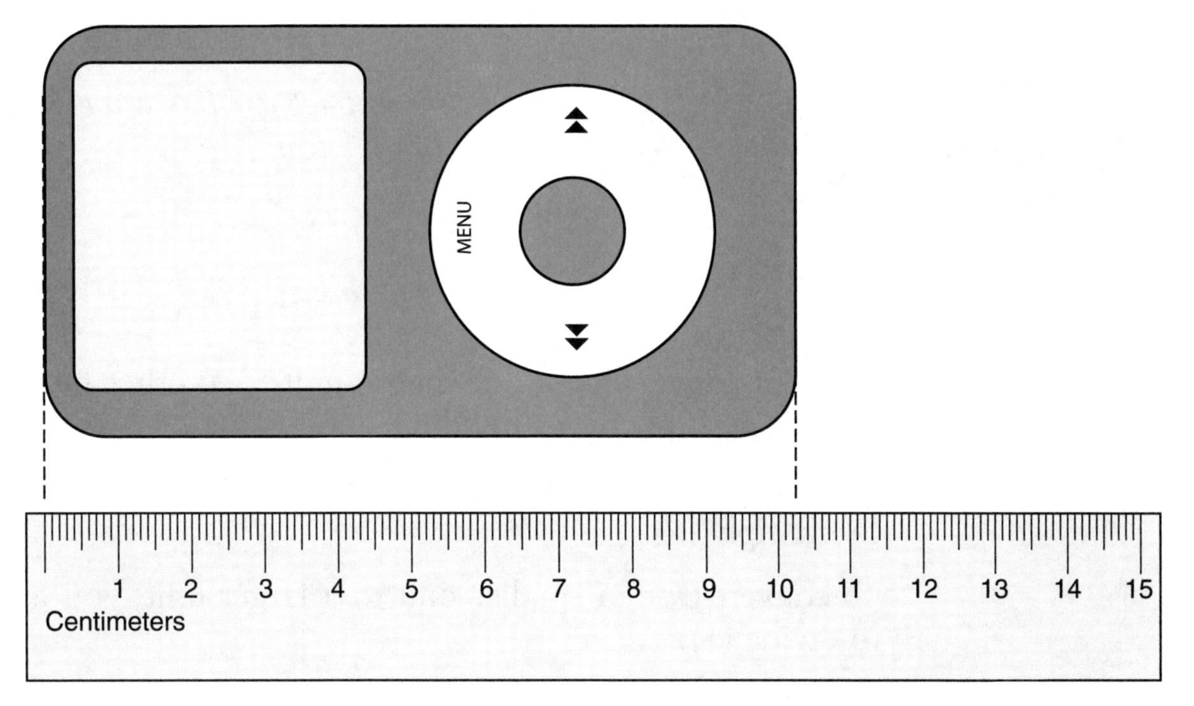

3. What is the length of this Ipod to the nearest centimeter? _____

Answers on page 179.

UNITS OF WEIGHT

The three most common U.S. customary units used to measure weight are **ounce (oz), pound (lb),** and **ton (T).**

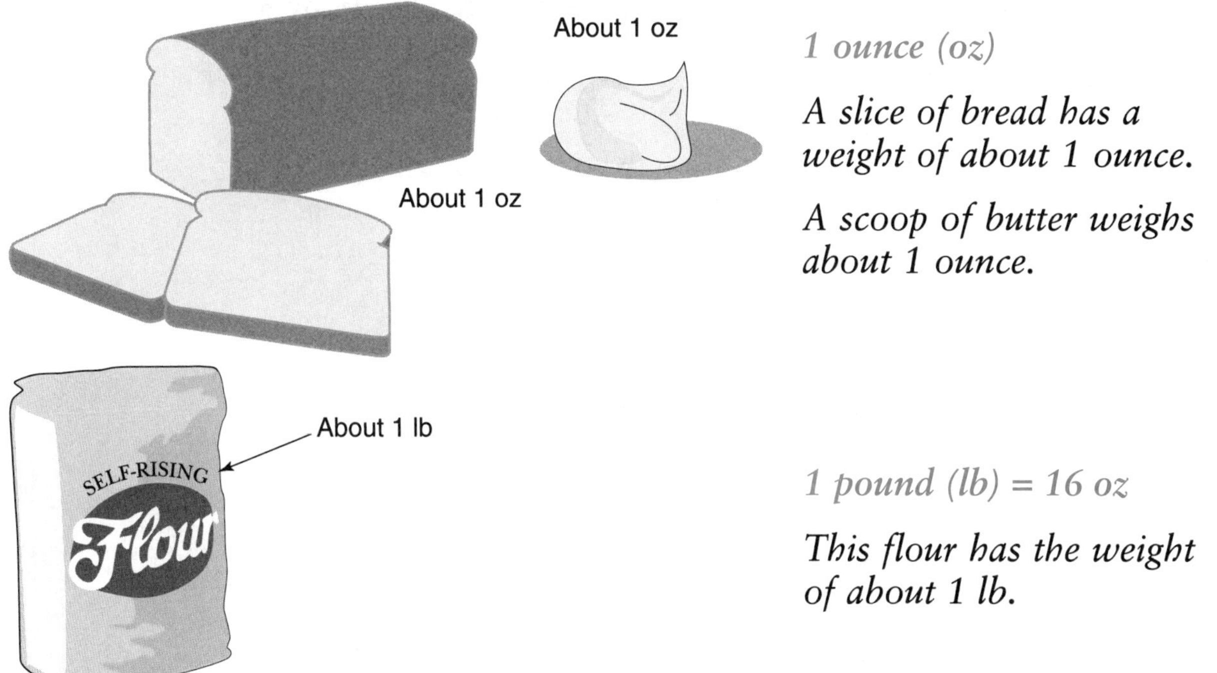

1 ounce (oz)

A slice of bread has a weight of about 1 ounce.

A scoop of butter weighs about 1 ounce.

1 pound (lb) = 16 oz

This flour has the weight of about 1 lb.

1 ton (T) = 2,000 lb

A compact car has a weight of about 1 ton.

CONVERTING THE U.S. TRADITIONAL WEIGHT

You can convert from one metric unit to another metric unit by multiplying or dividing.

To convert from a larger unit to a smaller unit, you have to multiply (×).

To convert from a smaller unit to a larger unit, you have to divide (÷).

Convert	Into	How to	Example
pounds	ounces	multiply by 16	3 lb = 48 oz
tons	pounds	multiply by 2,000	12 T = 24,000 lb

Do the opposite to convert back.

Example

To convert ounces into pounds divide by 16.

CONVERTING THE U.S. TRADITIONAL WEIGHT EXERCISES

1. 96 oz = _____ lb
2. 18,000 lb = _____ T
3. 3.5 lb = _____ oz
4. 16 lb = _____ oz
5. 64 oz = _____ lb
6. 20,000 lb = _____ T

Directions: Write the units of weight that would be best to measure the following.

7. weight of a bulldozer ______________

8. your weight ______________

9. weight of a cell phone ______________

10. weight of a cat ______________

Answers on page 179.

METRIC UNITS OF WEIGHT

The three common metric units used to measure weight are **milligram (mg), gram (g),** and **kilogram (kg).**

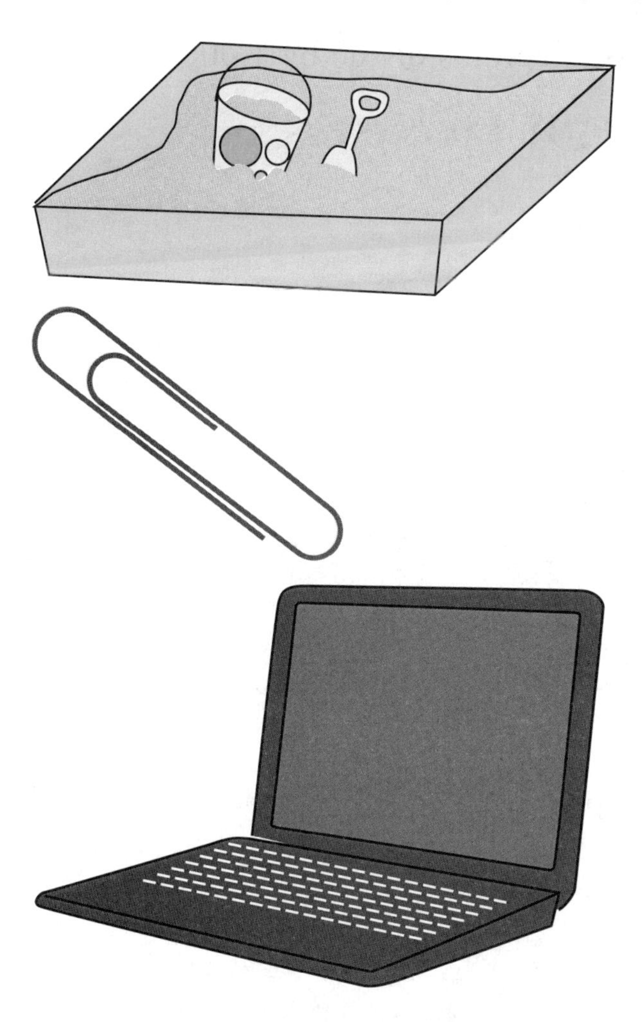

1 milligram (mg)

A single grain of sand has a weight of about 1 milligram.

1 gram (g)

A single paper clip has a weight of about 1 gram.

1 kilogram (kg)

A laptop has the weight of about 1 kilogram.

CONVERTING THE UNITS OF WEIGHT

You can convert from one metric unit to another metric unit by multiplying or dividing.

To convert from a smaller unit to a larger unit, you have to divide (÷).

To convert from a larger unit to a smaller unit, you have to multiply (×).

Convert	Into	How to	Example
grams	milligrams	multiply by 1,000	4 g = 4,000 mg
kilograms	grams	multiply by 1,000	12 kg = 12,000 g

Do the opposite to convert back.

Example

To convert grams into kilograms divide by 1,000.

CONVERTING THE UNITS OF WEIGHT EXERCISES

1. 3,200 g = _____ kg
2. 87 kg = _____ g
3. 6,100 mg = _____ g
4. 240 g = _____ kg
5. 64.2 g = _____ mg
6. 19,570 g = _____ kg

Directions: Write the units of weight that would be best to measure the following.

7. weight of a pineapple ______________
8. weight of a key ______________

9. weight of a calculator ______________

10. weight of a grain of salt _____________

Answers on pages 179 to 180.

UNITS OF VOLUME

The five most common units used to measure volume are **fluid ounces (fl oz), cup (c), pint (pt), quart (qt),** and **gallon (gal).**

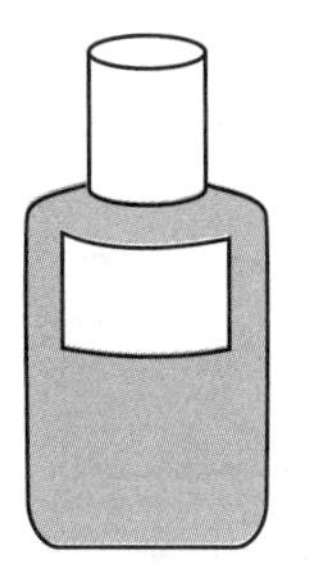

1 fluid ounce (fl oz)

A bottle of hand sanitizer has the volume of about 1 fluid ounce.

1 cup (c) = 8 fl oz

This juice pack holds 8 fluid ounces of orange juice.

1 pint (pt) = 2 c

A yogurt container like this one has the volume of about 1 pint.

1 quart (qt) = 2 pt

This juice container holds 1 quart.

1 gallon (gal) = 4 qt

A milk container like this one has a capacity of about 1 gallon.

CONVERTING VOLUME

You can convert one unit of volume to another by multiplying or dividing.

To convert from a smaller unit to a larger unit, you have to divide (÷).

To convert from a larger unit to a smaller unit, you have to multiply (×).

Convert	Into	How to	Example
cups	fluid ounces	multiply by 8	2 c = 16 fl oz
pints	cups	multiply by 2	5 pt = 10 c
quarts	pints	multiply by 2	6 qt = 12 pt
gallons	quarts	multiply by 4	4 gal = 16 qt

Do the opposite to convert back.

Example

To convert quarts into gallons divide by 4.

CONVERTING VOLUME EXERCISE

1. 9 c = _____ pt
2. 4 gal = _____ qt
3. 40 fl oz = _____ c
4. 4 c = _____ fl oz
5. 18 pt = _____ qt
6. 6 qt = _____ pt
7. 5 pt = _____ c
8. 36 qt = _____ gal

Answers on page 180.

METRIC UNITS OF VOLUME

The two other common metric units used to measure volume are **milliliter** (**mL**) and **liter** (**L**).

1 milliliter (mL)

This medicine dropper would hold 1 milliliter.

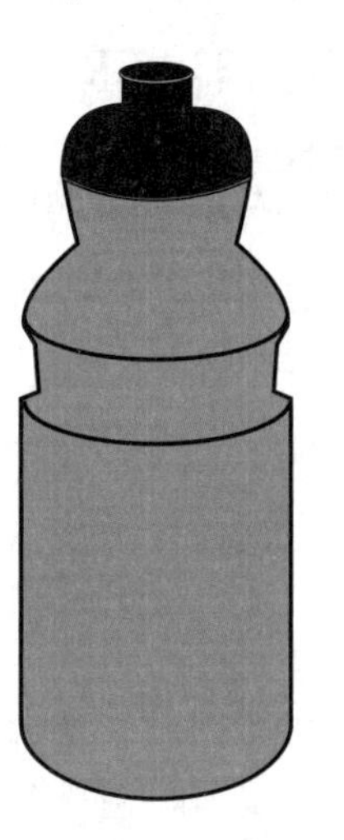

1 liter (L)

This water bottle holds about 1 liter.

CONVERTING THE METRIC VOLUME

You can convert from one metric unit to another metric unit by multiplying or dividing.

To convert from a smaller unit to a larger unit, you have to divide (÷).

To convert from a larger unit to a smaller unit, you have to multiply (×).

Convert	Into	How to	Example
milliliters	liters	divide by 1,000	23,000 mL = 23 L
liters	milliliters	multiply by 1,000	19 L = 19,000 mL

CONVERTING THE METRIC VOLUME EXERCISES

1. 9 L = _____ mL

2. 4,000 mL = _____ L

3. 3,500 mL = _____ L

4. 6 L = _____ mL

Answers on page 180.

PERIMETER

For the NJASK5, you only need to be familiar with finding the area and perimeter of squares and rectangles. The **Perimeter** *(P)* of a shape is the distance around the outside of a plane figure. Perimeter is measured in linear units. The formula for the perimeter of a rectangle is

$$P = (2 \times L) + (2 \times W) \quad \text{or} \quad P = L + W + L + W$$

Examples:

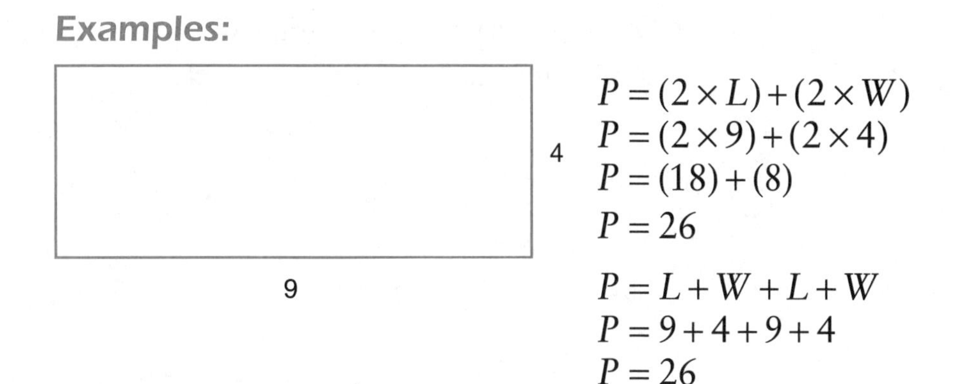

$$P = (2 \times L) + (2 \times W)$$
$$P = (2 \times 9) + (2 \times 4)$$
$$P = (18) + (8)$$
$$P = 26$$

$$P = L + W + L + W$$
$$P = 9 + 4 + 9 + 4$$
$$P = 26$$

The formula for the perimeter of a square is: $P = 4 \times s$

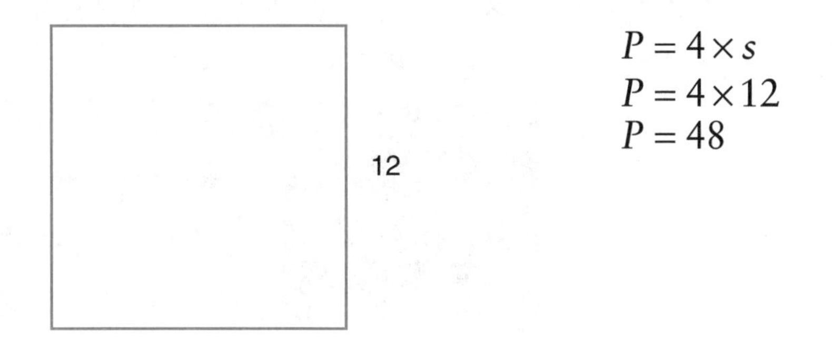

$$P = 4 \times s$$
$$P = 4 \times 12$$
$$P = 48$$

PERIMETER EXERCISES

Directions: Find the perimeter of each shape.

1.

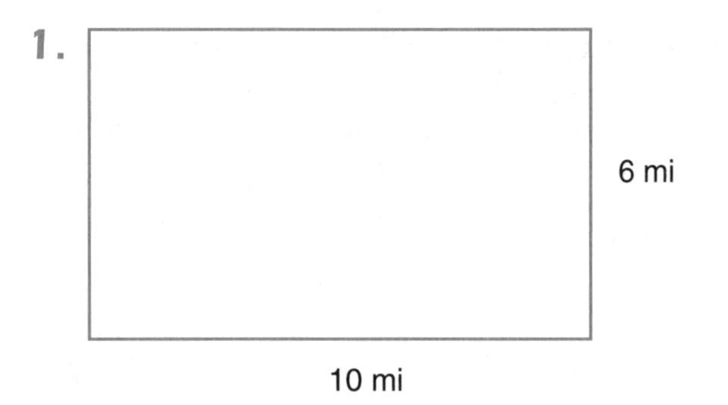

2. A regular polygon is below with side equal to 2 centimeters. Find the perimeter.

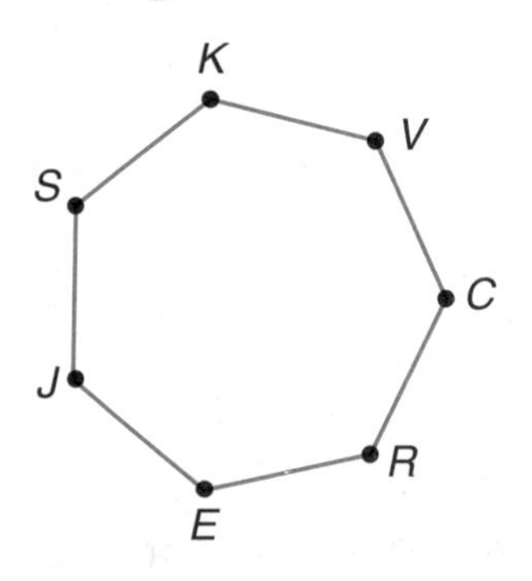

3. The flag below measures 2 feet by 6 inches. What is its perimeter in inches?

4. Find the perimeter of the shape below.

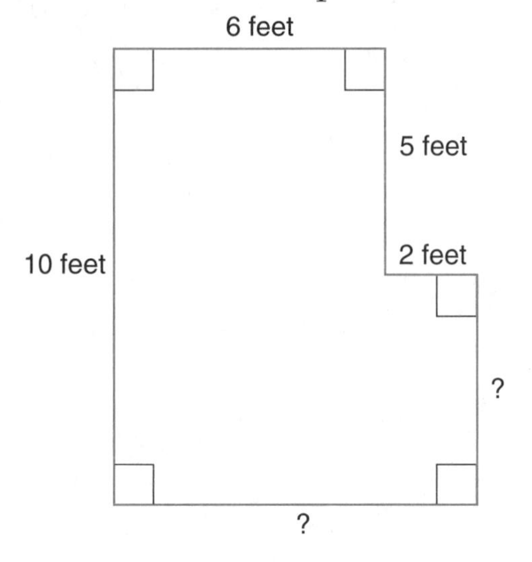

5. Mark's desk at school has the length of 3 ft and the width of 4 ft. What is the perimeter of Mark's desk at school?

6. Kathy has a rectangular mirror in her bathroom that has a length of 32 in. and a width of 18 in. What is the perimeter of Kathy's bathroom mirror?

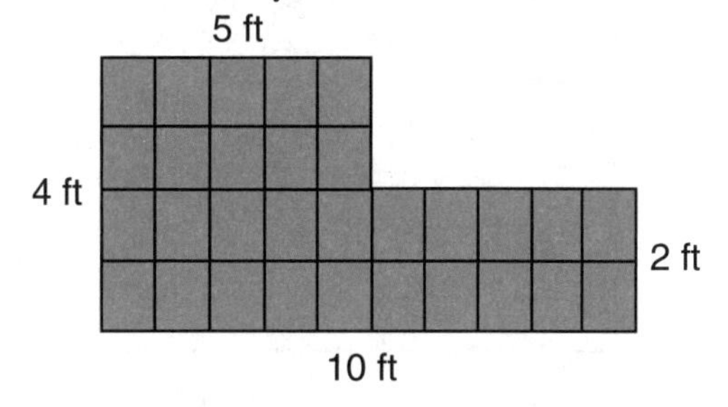

7. Given this figure above. Find its perimeter.

Answers on page 181.

AREA

The **Area** *(A)* is the amount of surface that a plane figure covers. Area is measured in square units. A square unit is a square that is 1 unit long and 1 unit wide.

Example 1.

Count the number of squares units in the rectangle.
The area of this rectangle is 40 square units.

The formula for the area of a rectangle is

$$A = L \times W$$

Example 2.

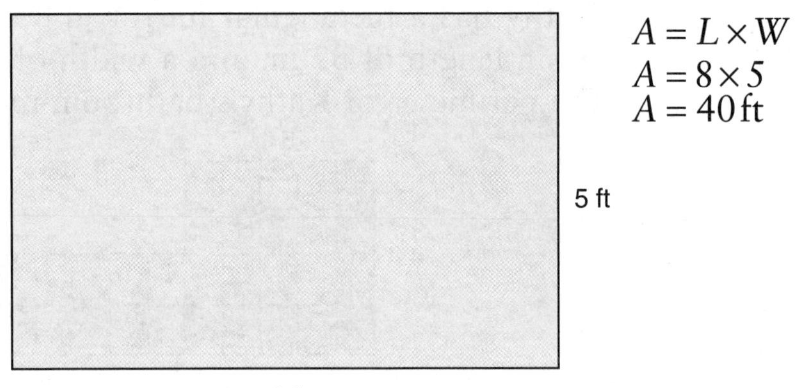

$A = L \times W$
$A = 8 \times 5$
$A = 40 \text{ ft}$

$A = 40$

Example 3.

$A = L \times W$
$A = 2 \times 2$
$A = 4 \text{cm}$

AREA EXERCISES

1. Find the area.

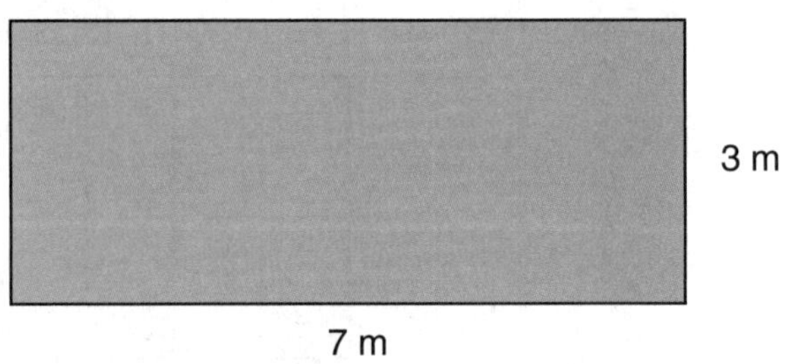

2. Find the perimeter and area of the square below.

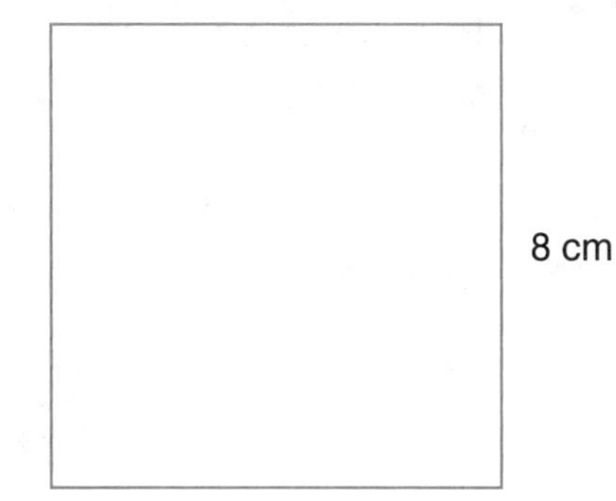

3. Find the area and perimeter of the picture below.

4. If a square has area = 225 square units, find the length of one side.

5. Which of the following describes a shape that has an area of 40 square centimeters?

   A. A rectangle with length of 10 centimeters and width of 5 centimeters

   B. A rectangle with length of 10 centimeters and width of 10 centimeters

   C. A square with length of 10 centimeters

   D. An isosceles triangle that has two sides 8 centimeters long and one side 4 centimeters long

6. The parking lot of the Cheeseburger Madness is the shape of a rectangle. The length is 24 yards and its width is 12 yards. The entire parking lot is being redone. How much area is going to be redone?

7. Billy's soccer team plays on a field that is 80 yards long and 55 yards wide. What is the area of the soccer field?

8. Given the playground below, determine its area in meters.

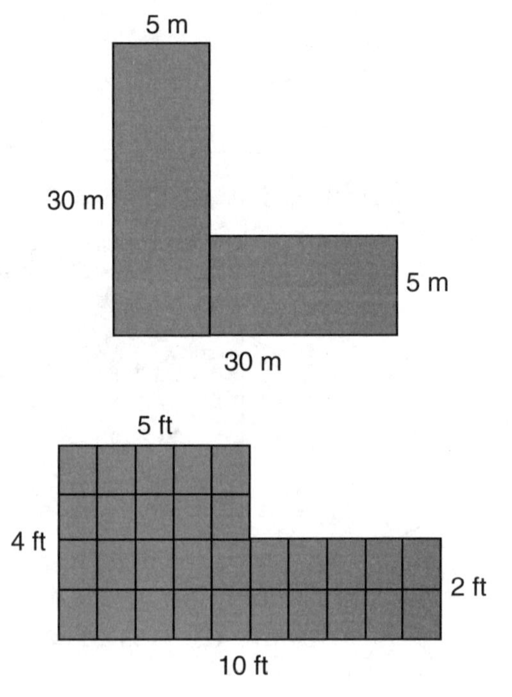

9. Given the figure above. Find its area.

10. The figure below is a kitchen floor. All of the square tiles on the floor need to be replaced EXCEPT for the shaded ones. How many square feet of tile need to be replaced?

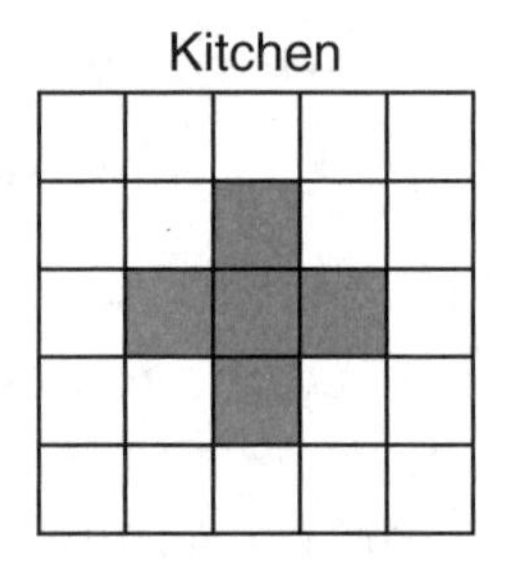

Answers on pages 181 to 182.

COORDINATE GRID

The coordinate grid is formed by two number lines that intersect at right angles at point (0, 0) or the point of origin. The horizontal number line is called the x-axis and the vertical number line is called the y-axis.

Locations of points on the plane can be plotted when one coordinate from each of the axes is used. This set of x and y values is called ordered pairs. Points on a grid are located by ordered pairs or coordinates. The first coordinate on a grid is the x-coordinate, which can be determined by traveling horizontally (right). The second coordinate is the y-coordinate, which can be determined by traveling vertically (up).

Example 1.

Starting at the point of origin (0, 0), the x-coordinate of this point is four units to the right. The y-coordinate of this point is five units up. The ordered pair or the (x, y) coordinates is (4, 5). Remember that the x-coordinate always comes first.

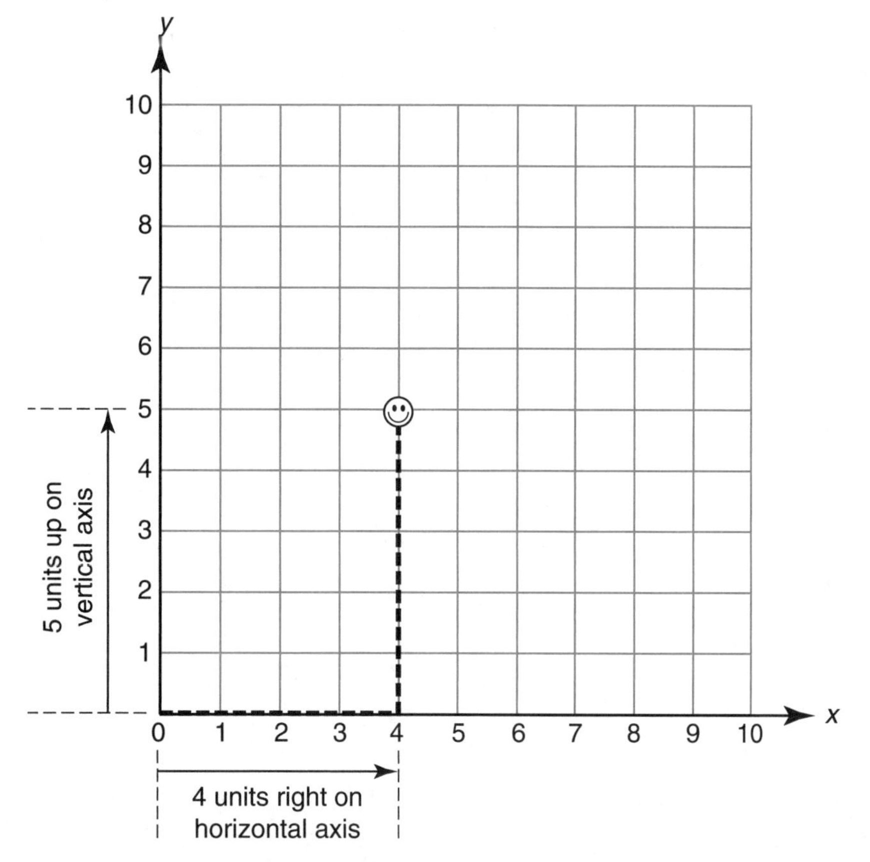

Example 2.

Starting at the point of origin (0, 0), the x-coordinate of this point is 0 units to the right. The y-coordinate of this point is 7 units up. The ordered pair or the (x, y) coordinates is (0, 7).

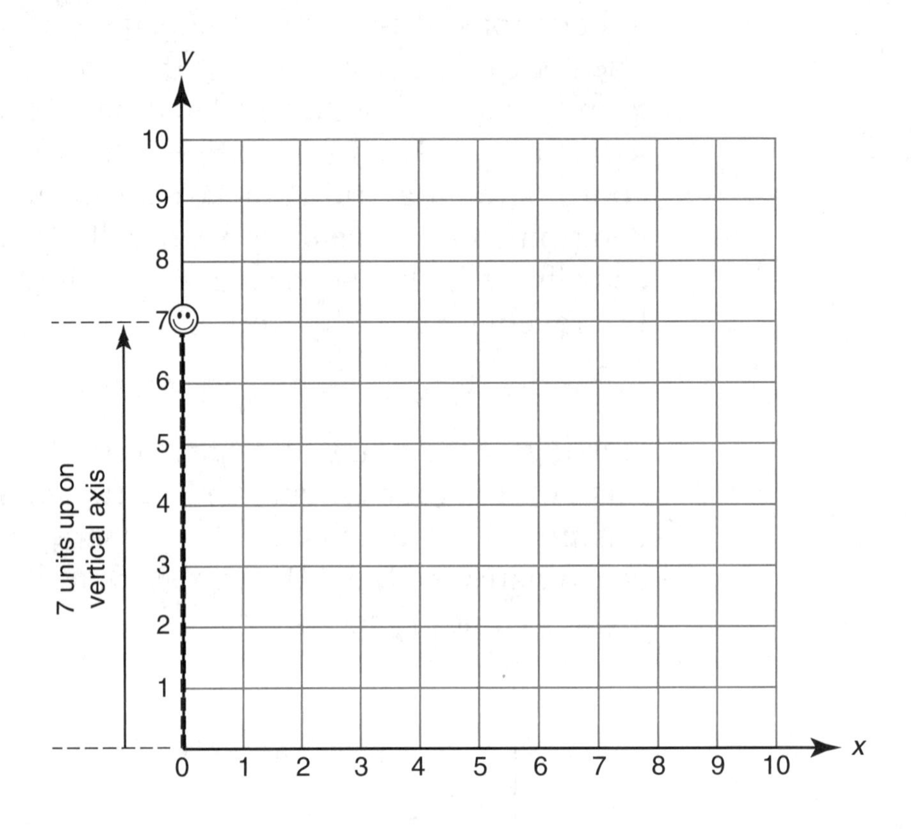

COORDINATE GRID EXERCISES

What are the coordinates of the following?

1. Smiley face _____
2. Star _____
3. Basketball _____
4. Hamburger _____
5. Soccer ball _____

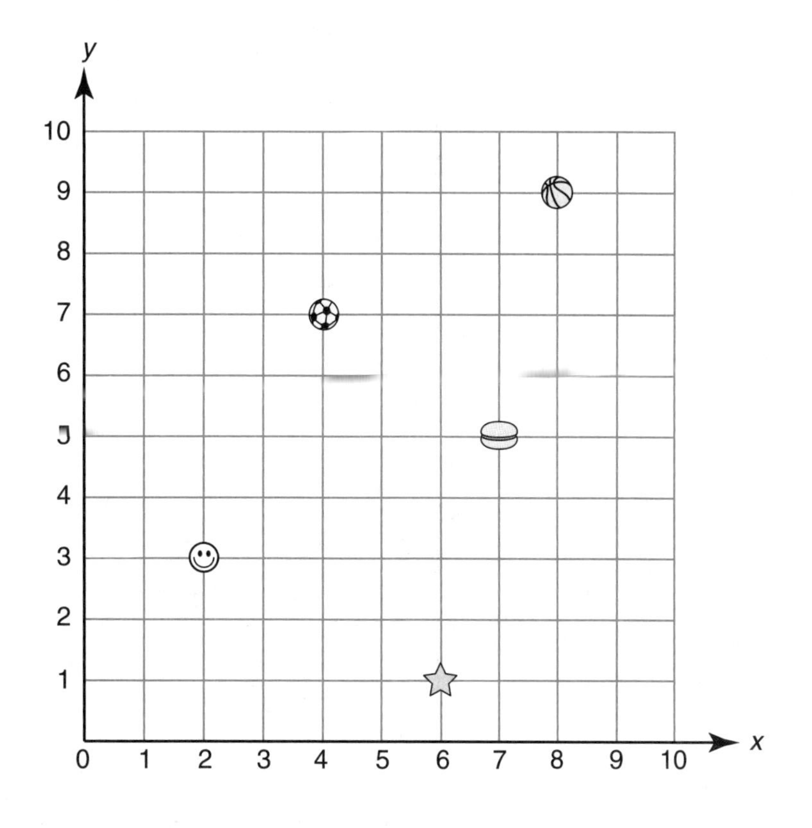

6. What object can be found at (6, 4)?

7. What object is located 4 units right and 6 units up?

8. What object can be found at (9, 1)?

9. What object is located 7 units right and 10 units up?

10. What object can be found at (1, 0)?

11. What object can be found at (8, 7)?

12. What object can be found at (3, 2)?

13. What object can be found at (2, 9)?

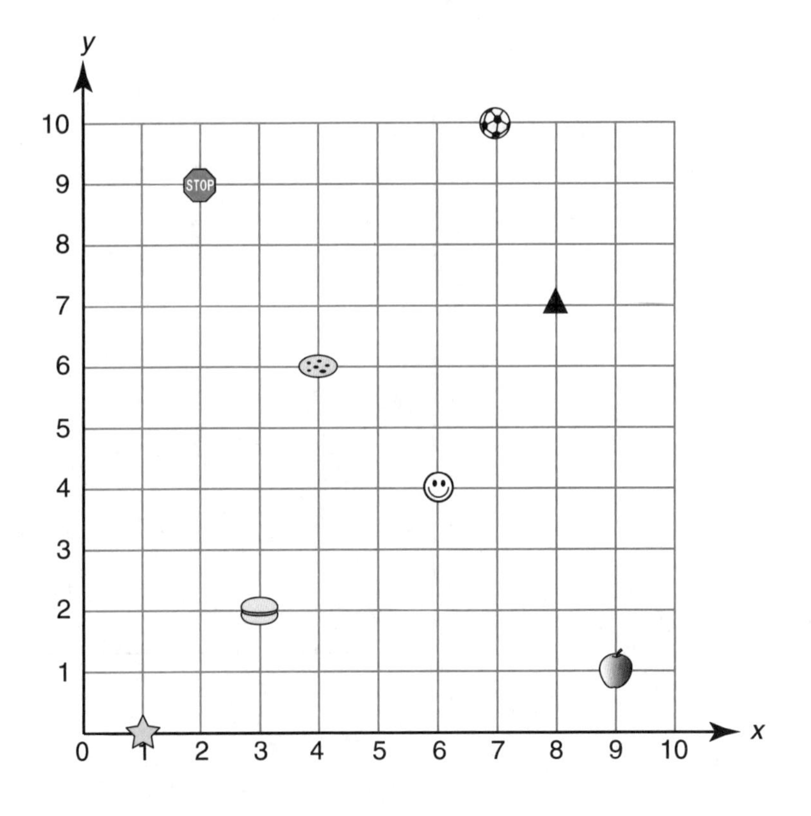

Point *M* is located at (3, 5). Use the coordinate grid below to plot the following coordinates.

14. *A* (7, 1)

15. *T* (2, 9)

16. *H* (5, 3)

17. *I* (6, 8)

18. *S* (0, 4)

19. *F* (10, 2)

20. *U* (9, 0)

21. *N* (8, 5)

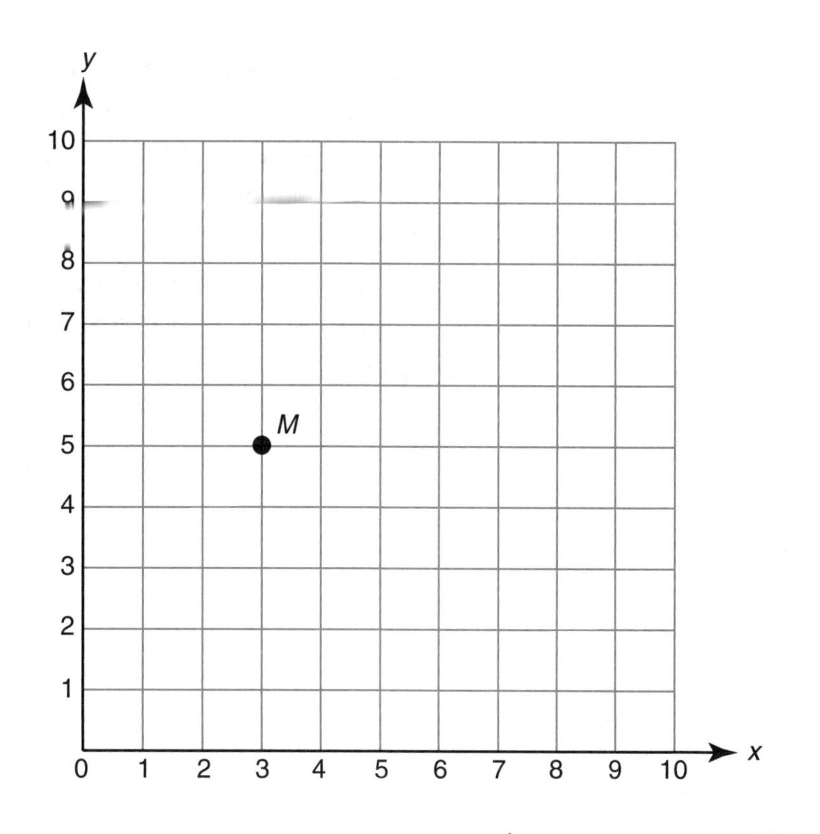

Answers on pages 182 to 183.

EXTENDED CONSTRUCTED-RESPONSE ITEMS

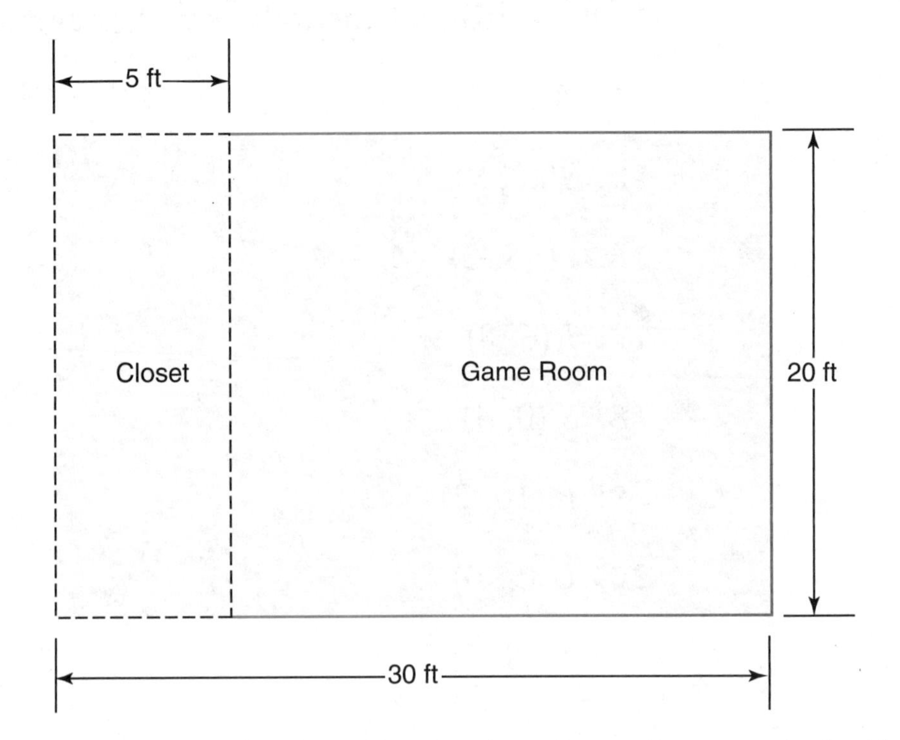

1. Eric is purchasing tiles for his Game Room. Each tile is one foot square and cost $2.00 each.
   - How much will it cost to tile the entire game room including the closet?
   - Eric decides to tile the Game Room without tiling the closet. How much will Eric save by tiling the Game Room only? Show your work and explain your answer.

Answers on page 183.

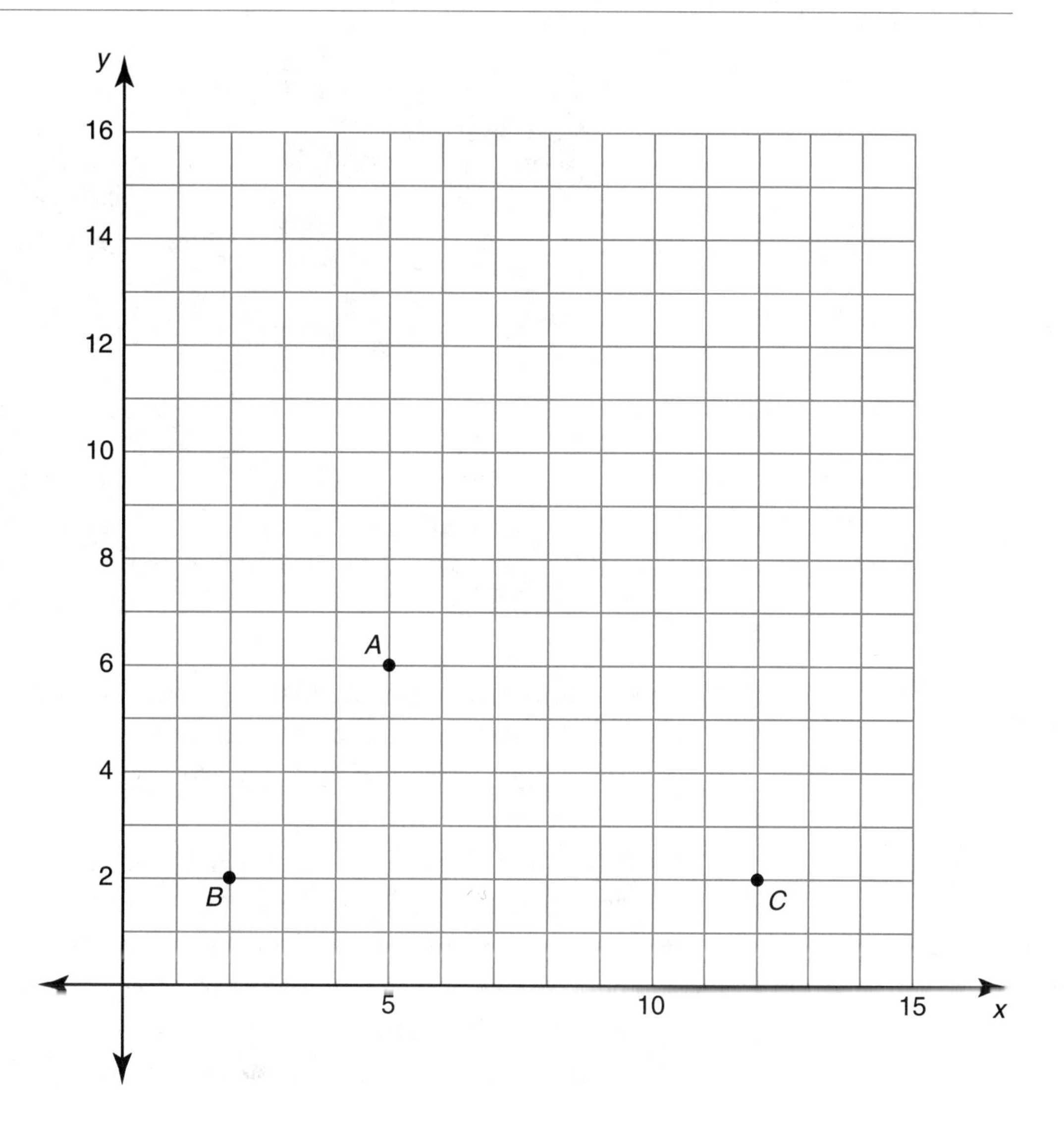

2. Three vertices of a parallelogram are shown above.
   - Complete the parallelogram by plotting the fourth vertex.
   - Label the fourth vertex point *D* and list the coordinates of all four vertices.
   - If the distance from vertex *A* to vertex *B* is 5 units, what is the perimeter of parallelogram *ABCD*? Show and explain work.

Answers on page 183.

3. Kesha loves this picture of her family vacation in Florida. She would like to create a wall mural of this picture. Kesha would like to enlarge all side lengths so that they are 9 times longer.
   - What are the dimensions of the enlarged picture in feet? Show work and explain answer.
   - She also plans to use line(s) of symmetry to center the picture. Draw in the line(s) of symmetry on the picture above to give Kesha an idea of how the enlarged picture might be centered.

Answers on page 184.

Chapter 5

PATTERNS AND ALGEBRA

All students will represent and analyze relationships among variable quantities and solve problems involving patterns, functions, and algebraic concepts and processes.

WHAT IS ALGEBRA?

- Algebra is a symbolic language used to express mathematical relationships. Students need to understand how quantities are related to one another, and how algebra can be used to concisely express and analyze those relationships.
- Algebra provides the language through which we communicate the patterns in mathematics. From the earliest age, students should be encouraged to investigate the patterns that they find in numbers, shapes, and expressions and, by doing so, to make mathematical discoveries. They should have opportunities to analyze, extend, and create a variety of patterns and to use pattern-based thinking to understand and represent mathematical and other real-world phenomena.
- Algebra is used to model real situations and answer questions about them. This use of algebra requires the ability to represent data in tables, pictures, graphs, equations, and rules. Modeling ranges from writing simple number sentences to help solve story problems in the primary grades to using functions to describe the relationship between two variables, such as the height of a pitched ball over time.

- The algebraic function concept is one of the most fundamental unifying ideas of modern mathematics. Students begin their study of functions in the primary grades, as they observe and study patterns. As students grow and their ability to abstract matures, students form rules, display information in a table or chart, and write equations that express the relationships they have observed.
- Algebra is a gatekeeper for the future study of mathematics, science, the social sciences, business, and a host of other areas. It is important that algebra play a major role in a mathematics program that opens the gates for all students.

PATTERNS

Patterns are part of our everyday lives. Wherever we look, we see patterns. There are patterns in nature, in textiles, and in carpets and tiles. There are patterns in numbers, words, colors, and music. Breaking down patterns and creating or identifying patterns help children begin to break down codes in more than just mathematics.

The above sequence of circles and hearts is a pattern. Children at a young age should experience such patterns. The one above is commonly referred to as an AB pattern as it has only two parts: A = circle and B = heart.

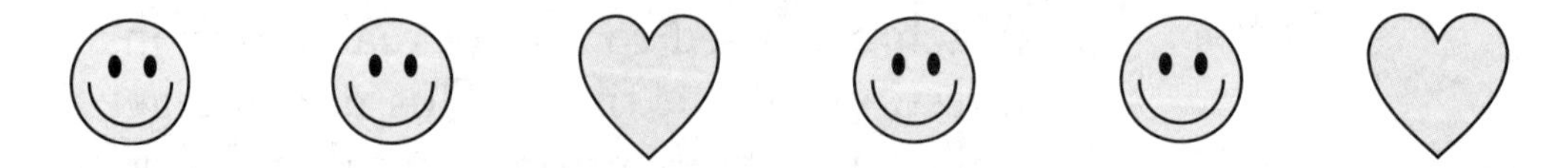

The new pattern above is an AAB pattern.
Arithmetic patterns are common in the elementary grades.

The simplest of these is 1, 2, 3, 4, 5, . . .

Others are the even numbers 2, 4, 6, . . . and the odd numbers 1, 3, 5, . . .

By fifth grade, it is expected that students can look for patterns that involve more than one operation, count backwards, or use missing numbers in the beginning or middle of the pattern (rather than the end). Sometimes students may be given the pattern and be asked to describe the rule. A few more examples follow below:

30, 27, 24, 21, . . . (counting back by 3s)

40, __, 30, 25, __, 15, 10 (counting back by 5s)

1, 2, 4, 5, 7, 8, 10 (adding one, adding two, adding one, adding two)

Example

A pattern can also be illustrated as below. The top of the pyramid has 1 block, the second layer has 4 blocks, the third layer has 9 blocks, and the fourth layer has 16 blocks. If a new layer was added at the bottom, how many blocks would be needed?

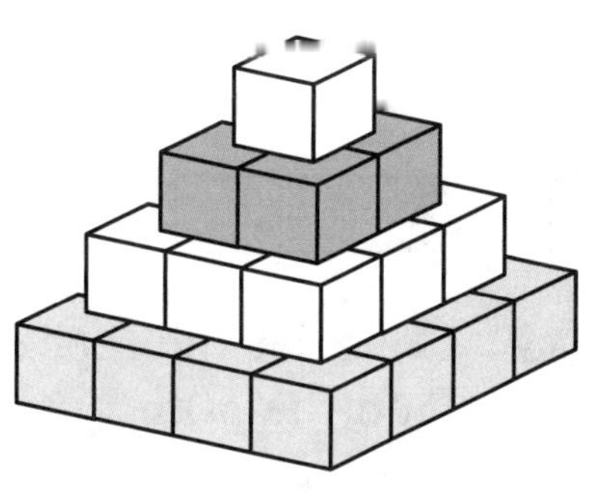

The next layer would have 25 blocks in a 5 by 5 square arrangement. Note that each layer is a square (1 by 1, 2 by 2, 3 by 3, . . .).

PATTERNS EXERCISES

1. Draw an ABB pattern.

2. Draw an ABC pattern.

3. What type of pattern is pictured below? Draw the next ten blocks.

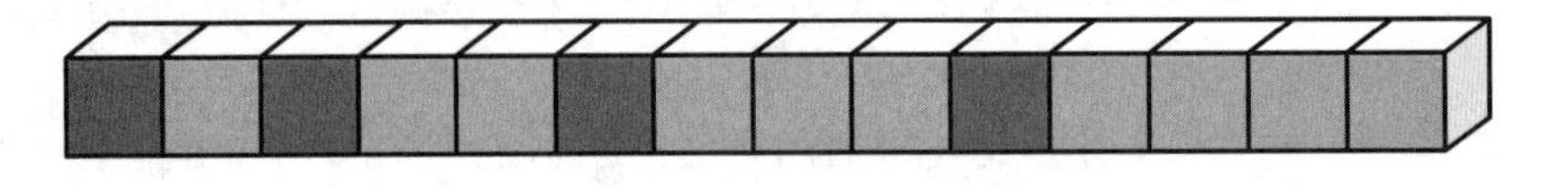

Questions 4–11—Find the next or missing items in the pattern.

4. 1, 2, 1, 1, 2, 1, 1, 1, 2, . . .

5. 2, 5, _____, 11, 14, _____, 20

6. 1, 2, 4, 8, . . .

7. Ann, Brad, Carol,

8. 16, 37, 58, . . .

9. 1, 1, 2, 3, 5, 8, 13, . . .

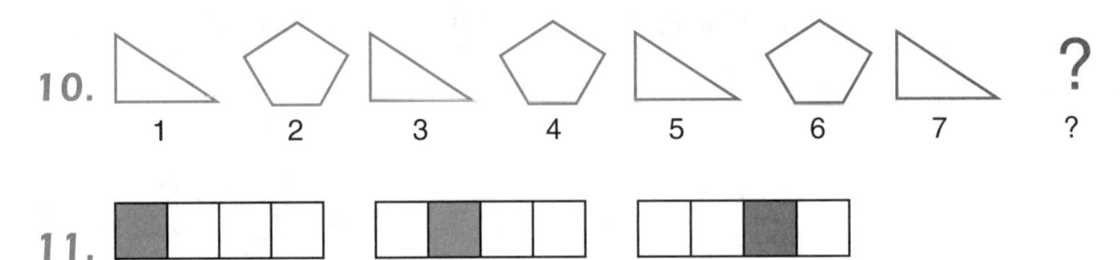

12. Consider the repeating ABCD pattern: 1, 4, 2, 8. Find the 50th number in this pattern.

13. 1, 1, 2, 4, 6, 18, 20, . . .

14. The temperature outside was 70°F at 5 p.m. If the temperature went down 3°F every hour, what was the temperature outside at 11 p.m.?

15. Find the next three letters in the pattern. O, T, T, F, F, S, . . .

16. Complete the sequence 1, 2, 4, . . . in three different ways. Explain your solutions.

Answers on pages 184 to 185.

FUNCTIONS

Functions are rules that describe the relationship between two things or variables. The two variables in a function have names that explain how they relate to one another. One variable, the input (x), is known as the independent variable. This is the variable upon which the operation or operations are performed. The other variable, the output (y), is called the dependent variable. The value of this variable is dependent upon the value of the independent variable.

Many functions are displayed as tables. For example, consider the table below describing pencils and their cost. This table displays a functional relationship between the variables Pencils (P) and their Cost (C). For each value of P, we have a cost C. That is, if we know what P is, then we can find C. This is the essence of a function: knowing one variable's value enables us to find the corresponding value of the other variable. A function table is a table of ordered pairs following the rule for that function.

Number of Pencils	Cost
1	7¢
2	14¢
3	21¢
4	28¢
10	70¢

Example 1

How much would 15 pencils cost?
The table shows that each pencil costs 7¢; thus 15 pencils cost 15 × 7¢ = \$1.05.

Example 2

How many pencils could you buy for \$1.40?

$$140/7 = 20 \text{ pencils}$$

Example 3

What is the relationship that describes the function?

Cost = 7¢ times (the number of pencils)

Often students will be told that the pattern in the table will continue and they will have to find "other" values. The function table below shows the number of pull-ups that Charisa has done in gym class each week. If the pattern continues, how many pull-ups will Charisa do in the tenth week?

Week	Pull-ups
1	4
2	7
3	11
4	16
5	22

Students can extend the table following the pattern to arrive at the answer of 67 pull-ups.

Week	Pull-ups	
1	4	
2	7	3 more
3	11	4 more
4	16	5 more
5	22	6 more
6	29	7 more
7	37	8 more
8	46	9 more
9	56	10 more
10	67	11 more

Functions can also be displayed or described in the form of a graph. A table describing the cost of parking per car is shown next. The graph shows the points from the table.

Number of Cars	Cost
1	3
2	6
3	9
4	12
5	15

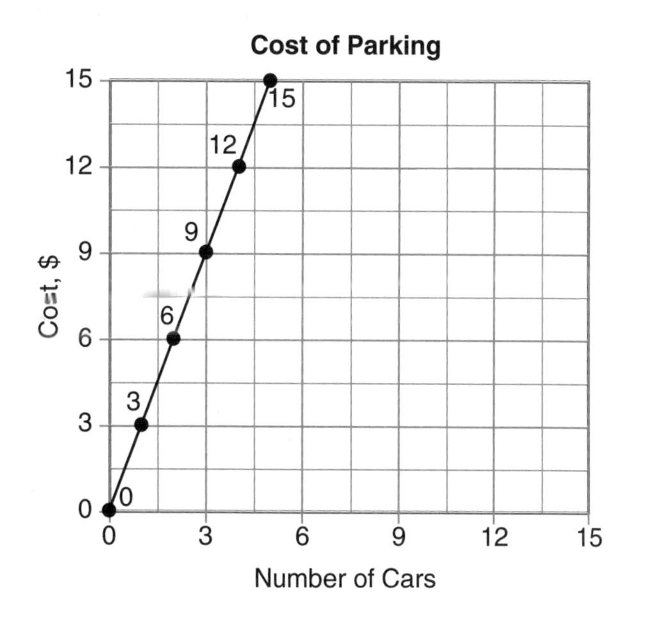

It is crucial to note the following:

- The table values can be grouped in pairs (1, 3), (2, 6), (3, 9), (4, 12), and (5, 15). These are commonly referred to as ordered pairs.
- The first numbers in the ordered pair (the independent variable) are graphed on the horizontal axis (x-axis).
- The second numbers in the ordered pair (the dependent variable) are graphed on the vertical axis (y-axis).

Example 4

The school cafeteria sells hot dogs. Each hot dog costs $2. Complete the table below for sales of 1, 2, 3, 4, and 5 hot dogs and graph the data on the coordinate grid.

Number of hot dogs sold	Cost
1	
2	
3	
4	
5	

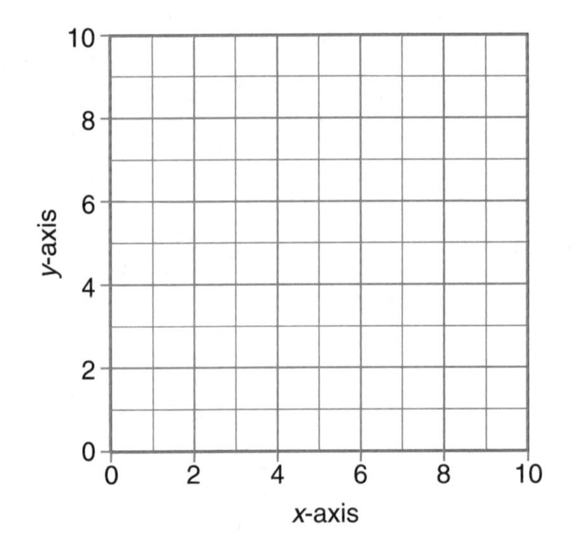

Example 5

A diagram called a function machine is sometimes used instead of a table. The pictures below show two numbers entering the machines (inputs) and two numbers exiting the machine. What does the machine do to the numbers?

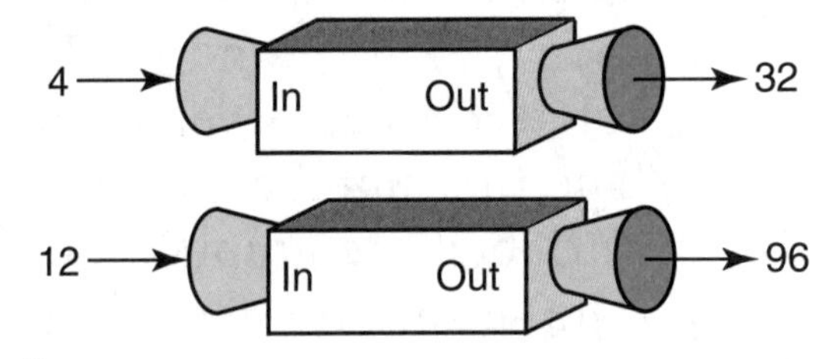

Multiply by 8.

FUNCTIONS EXERCISES

1. Find the missing numbers in the table below.

In	Out
2	4
3	6
11	22
27	?
?	18

2. Find the missing numbers in the table below.

In	Out
2	7
4	13
7	22
10	31
12	?
?	76

3. Find the missing words in the table below.

In	Out
One-eyed monster	3
Three-eyed monster	11
Four-eyed monster	15
?	7

4. Find the missing words in the table below.

In	Out
House	4
Cup	2
Writer	5
Mathematics	?
Elephant	?
?	8
?	0

5. Toothpicks were used to make the pattern below.

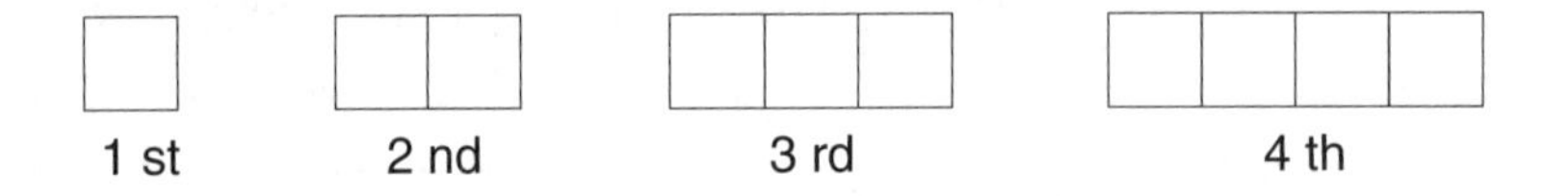

- How many toothpicks will be in the fifth figure? in the sixth figure?
- Write a table for the number of toothpicks needed to make the *n*th figure.
- Identify and describe the figure in this pattern that can be made with exactly 100 toothpicks.
- Make a graph of the data.

6. Toothpicks were used to make the pattern below.

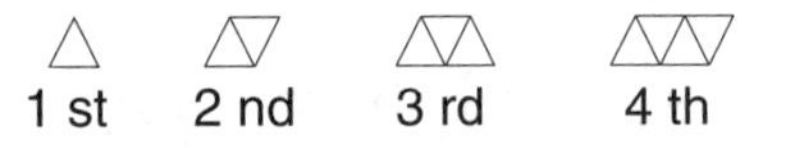

- How many toothpicks will be in the fifth figure? in the sixth figure?
- Write a table for the number of toothpicks needed to make the *n*th figure.
- Identify and describe the figure in this pattern that can be made with exactly 101 toothpicks.
- Make a graph of the data.

7. Using the table below, find the output if the input is 59. Write a rule.

In	10	17	28	40
Out	50	43	32	20

8. Tyrone has $20 to buy as many hamburgers as he can for his friends. If the pattern in the table continues, what is the maximum number of hamburgers that Tyrone can buy? Graph the data.

Number of hamburgers	1	2	3	4
Cost	$1.50	$3.00	$4.50	$6.00

9. Below is pictured a function machine. Using this machine, what is the output, if the input is 10?

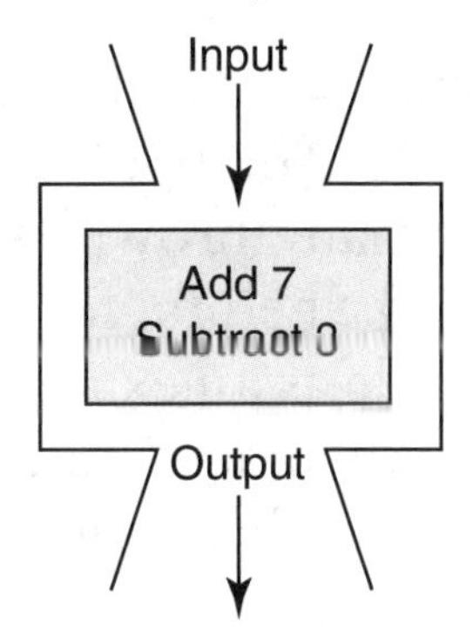

10. Which information matches the pattern shown in the following graph?

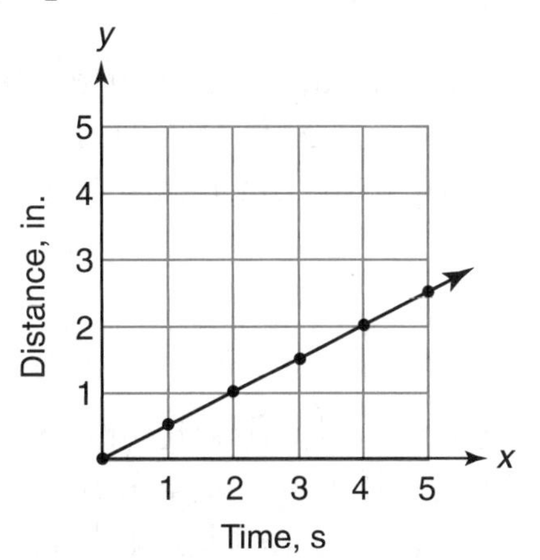

A. A mouse can move 2 inches in 1 second.
B. A snake can slither 2 inches in 2 seconds.
C. A fish can swim 2 inches in 4 seconds.
D. A turtle can crawl 2 inches in 3 seconds.

Answers on pages 185 to 188.

NUMBER SENTENCES AND VARIABLES

A number sentence is an equation or an inequality that involves numbers or variables. These problems are often the most difficult questions on the fifth grade NJASK. Often, a contextual sentence is given and students must translate it into a sentence. Four similar examples are given below:

Example 1

Patrice bought just enough nuts to put five on each brownie she made. If N is the number of nuts she bought, how many brownies did she make?
It can be helpful for students to select a number for the variable as an example. For instance, if Patrice bought 20 nuts and placed 5 nuts on each brownie, then she made $20 \div 5 = 4$ brownies. Thus, the correct number sentence would be

$$\text{Number of brownies} = N \div 5$$

Example 2

Lana bought five of each kind of cookies that a bakery made. If K is the number of kinds of cookies the bakery had, how many cookies did Lana buy?

$$\text{Number of cookies} = K \times 5$$

Example 3

Susan sold five fewer boxes of Girl Scout cookies than Lisa. If L is the number of boxes Lisa sold, how many boxes of cookies did Susan sell?

$$\text{Susan} = K - 5$$

Example 4

Harry brought 5 new packs of baseball cards today. If P is the number of packs he had yesterday, how many does he have now?

$$\text{Today} = P + 5$$

Number sentences can also have no context as below.

Example 5

If C represents a number, which of the following means "5 less than a number?"

$$C - 5$$

Number sentences can also be combined with a table as shown below.

Example 6

Which number sentence is true for all pairs of values shown in the table below?

Input	A	10	17	28	40
Output	B	50	43	32	20

$$A + B = 60$$

A number sentence could also involve an inequality (⟨or⟩) as the example below illustrates.

Example 7

Jacob and Issac each had $10.00 to spend for lunch. Jacob bought a meal that cost $8.50 and Issac bought lunch that cost $9.25. Write an inequality that compares the amount of money each student has left?

$$\$10.00 - \$9.25 < \$10.00 - \$8.50$$

A balance scale can be used to model how to find the missing number in an addition equation. The scale is balanced when the amount on the left side is equal to the amount on the right side of the balance scale.

Example 8

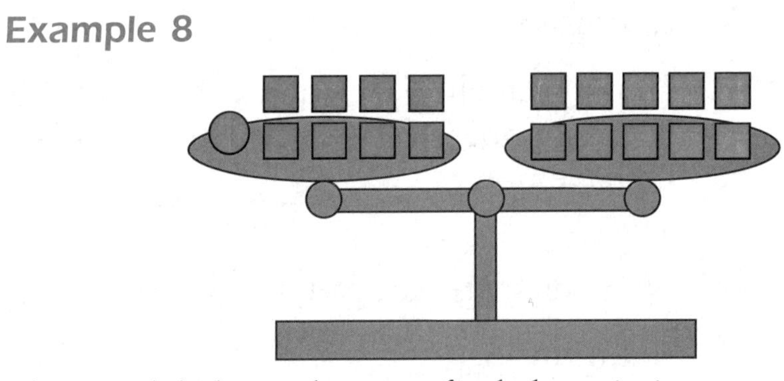

This model shows how to find the missing number in the equation $8 + x = 10$.

Each square is equal to 1 unit and the circle is the unknown (x).

Cross out 8 squares from each side of the model.

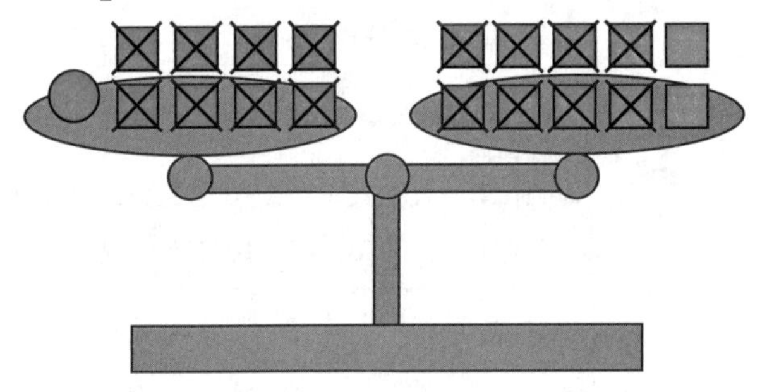

There are 2 squares left on the right side to balance the 1 circle on the left side.

Since the circle represents the unknown, $x = 2$.

NUMBER SENTENCES AND VARIABLES EXERCISES

1. There are 2 more drummers in the band than flute players. If F is the number of flute players in the band, how many drummers are there?
2. There are 2 fewer trumpet players in the band than flute players. If F is the number of flute players in the band, how many trumpet players are there?
3. The drummers in the band stand in the last 3 rows. The same number of drummers are in each row. If D is the total number of drummers in the band, how many sit in each row?
4. Tracy is packing 100 snacks for a class party. She packs 5 snacks in each bag. Write a number sentence that she can use to find the number of bags she will need.

5. Tanya is packing books into boxes. Each box can hold 15 books. Which number sentence can be used to find the total number of boxes that she needs in order to pack 75 books?

6. Mrs. Krulik is 65 years old. Her daughter is 30 years old. Mrs. Krulik's husband is k years old. Altogether, Mrs. Krulik's, her husband's, and her daughter's ages total 155. Write a number sentence that represents all of their ages.

7. If $t = 5$, then what does the expression $t + t + t - 3$ equal?

8. If $54 \div n = 6$, then what is the value of n?

9. What is the value of $n - 20$ when $n = 20$?

10. Start with any number, n. Determine which of the following will not result in the number n that you started with.

    A. Add 5, divide by 5

    B. Multiply by 5, divide by 5

    C. Divide by 5, multiply by 5

    D. Subtract 5, add 5

    E. Multiply by 5, add 5

11. Eric's age is E. Write a number sentence for each person's age below.

    A. Susan is five years older than Eric.

    B. Maria is twice as old as Eric.

    C. Donald is two years younger than Eric.

12. Kelly charges \$6 for each driveway she shovels. Last weekend, she earned \$30 shoveling driveways.

    A. Write a number sentence using s to represent the number of driveways shoveled.

    B. How many driveways did she shovel?

13. Using the model below, write a number sentence and solve for the unknown variable x.

Answers on pages 188 to 189.

GRAPHING LINEAR EQUATIONS FROM TABLES

Number patterns can be shown in graphs. The graph gives students an easy visual way to solve problems and to make further predictions based on the patterns seen in the graph.

Example 1

Graph the function $Y = X + 2$.

On a coordinate plane, the X values are on the horizontal axis. The Y values are on the vertical axis. Create a table by inserting values for X and Y. The table contains ordered pairs. Each X and Y pair must satisfy the equation $Y = X + 2$. Plug in 1 for the X value and then add 2 to get the Y value. Use the ordered pairs from the table to graph the line.

X	Y
1	3
2	4
3	5
4	6

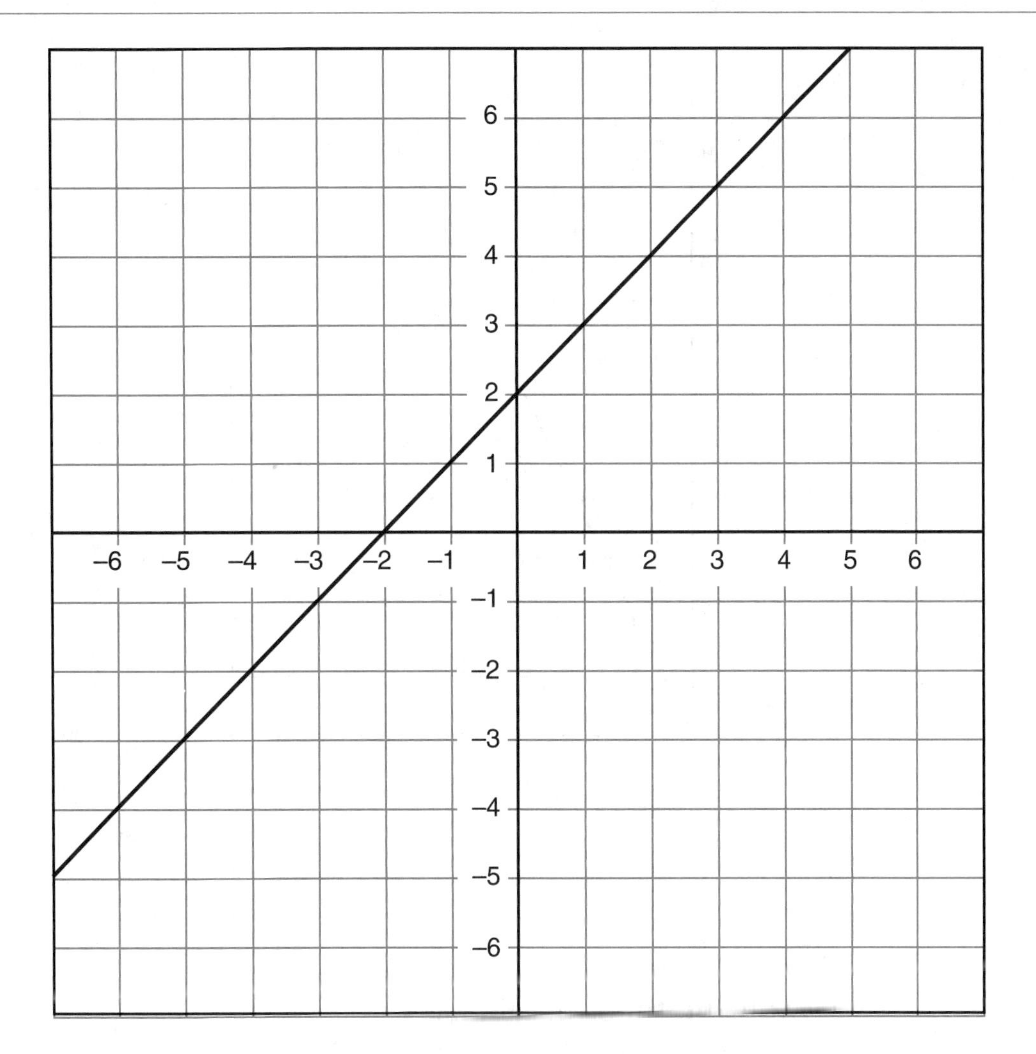

Example 2

Graph the function $Y = 2X$.

Create a table by inserting values for X and Y. The table contains ordered pairs. Each X and Y pair must satisfy the equation $Y = 2X$. The $2X$ means 2 times the X value. Use the ordered pairs from the table to graph the line.

X	*Y*
1	2
2	4
3	6
4	8

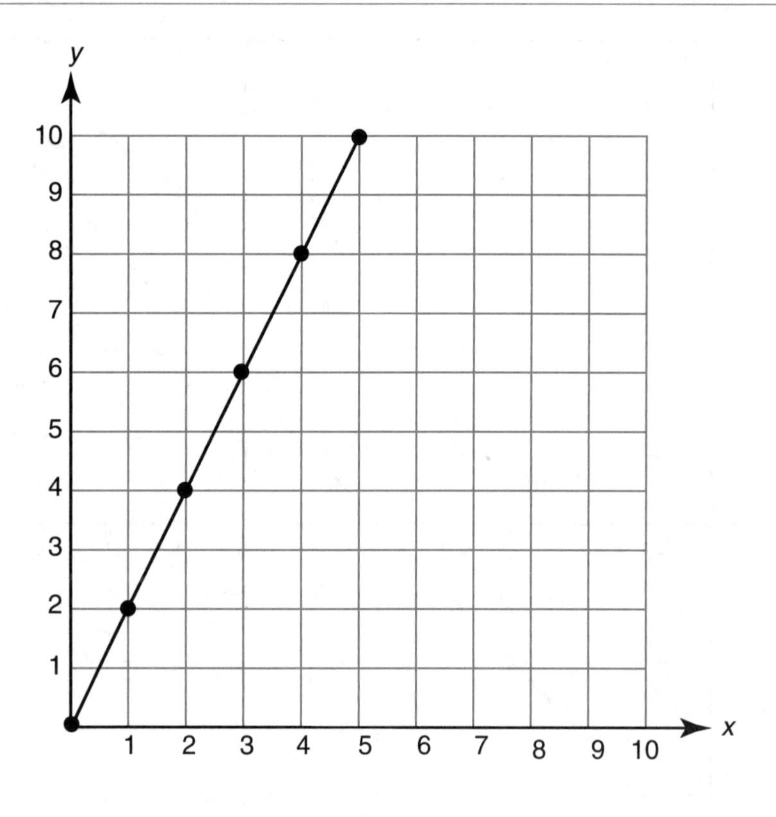

Example 3

Graph the function $Y = X - 1$.

Create a table by inserting values for X and Y. The table contains ordered pairs. Each X and Y pair must satisfy the equation $Y = X - 1$. Plug in 1 for the X value and then subtract 1 to get the Y value. Use the ordered pairs from the table to graph the line.

X	Y
1	0
2	1
3	2
4	3

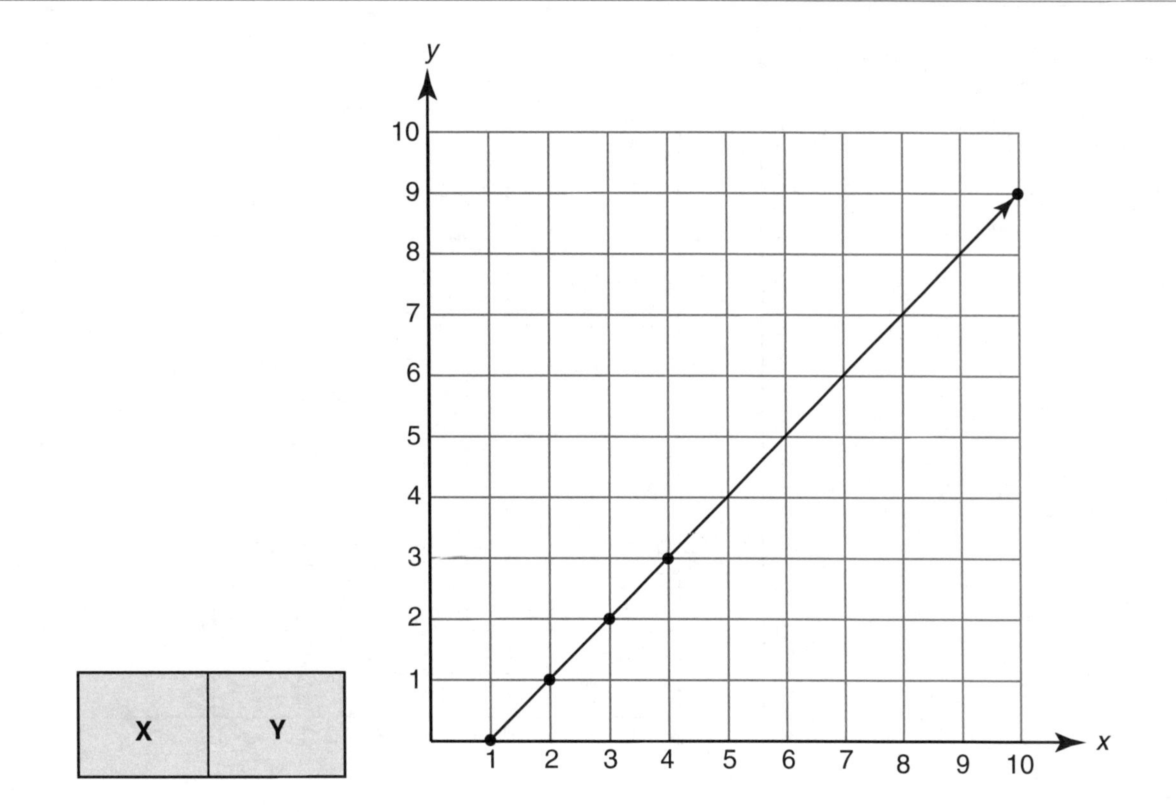

GRAPHING LINEAR EQUATIONS FROM TABLES EXERCISES

1. Complete the table with three ordered pairs that satisfy the equation $Y = X + 3$.

X	Y

2. Use the coordinates in your table to graph the line that represents the equation $Y = X + 3$.

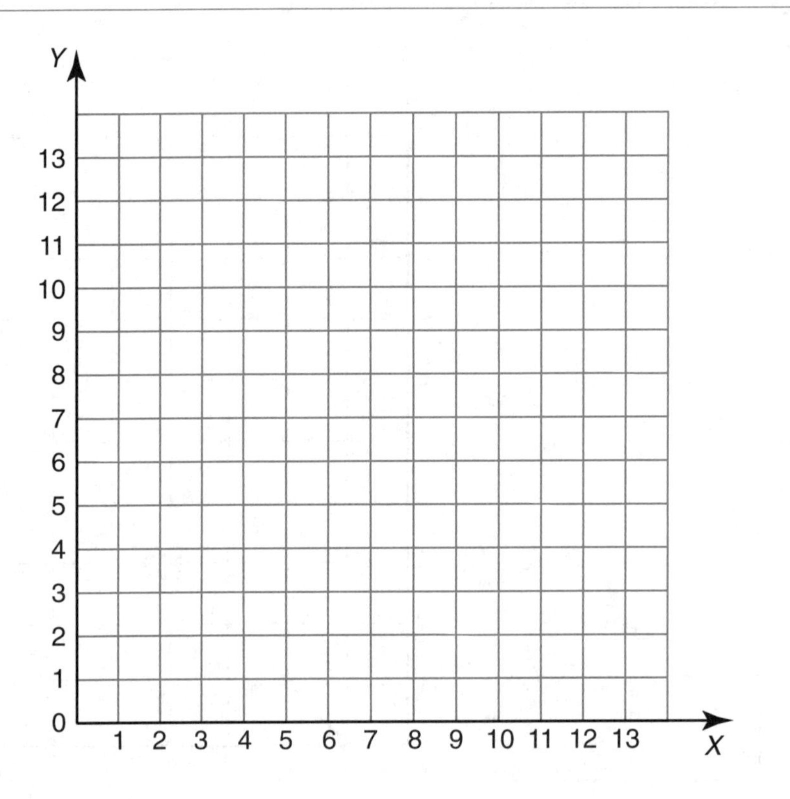

3. The line in the graph contains points (2, 3) and (3, 4). What equation is graphed below?

A. $Y = X + 2$

B. $Y = X + 1$

C. $Y = 2X$

D. $Y = X - 1$

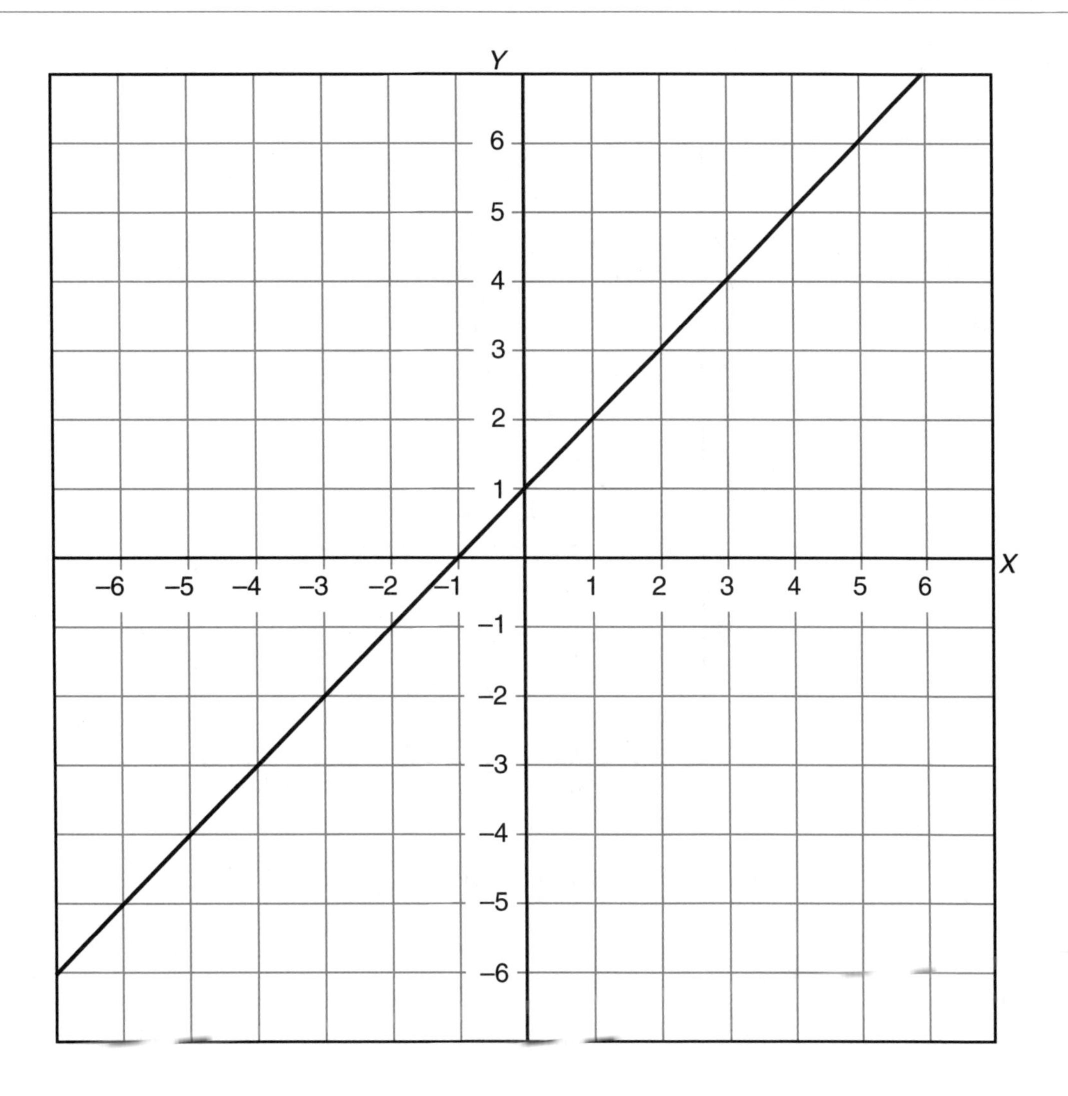
Y
6
5
4
3
2
1
X
−6 −5 −4 −3 −2 −1
1 2 3 4 5 6
−1
−2
−3
−4
−5
−6

4.

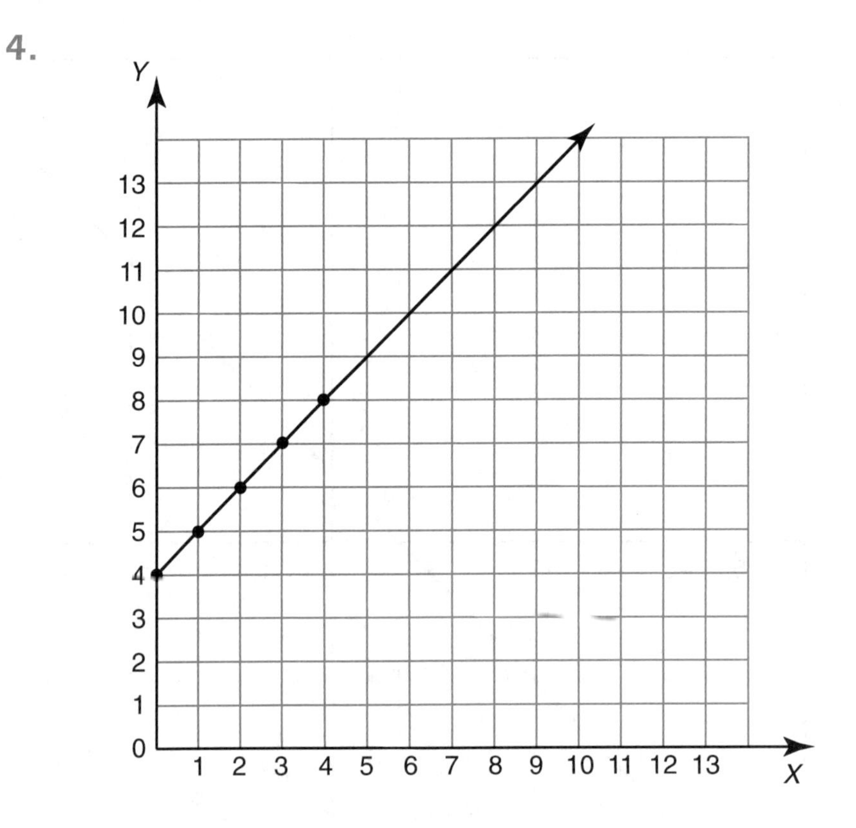

Which table shows the graph above?

A.

X	0	2	4
Y	2	2	6

B.

X	0	2	4
Y	3	5	4

C.

X	0	2	4
Y	4	6	8

D.

X	0	2	4
Y	5	7	9

5.

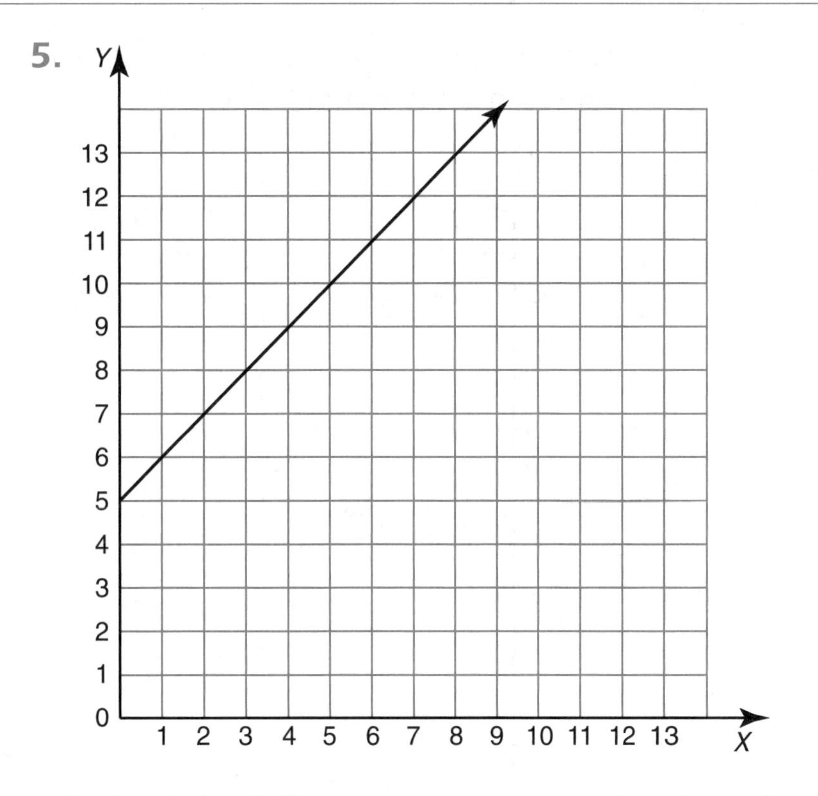

Which of the following points is on the above line?

A. (1, 5)

B. (4, 10)

C. (2, 7)

D. (8, 3)

6. Complete the table with three ordered pairs that satisfy the equation $Y = 4X$.

X	Y

7. Use the coordinates in your table to graph the line that represents the equation $Y = 4X$.

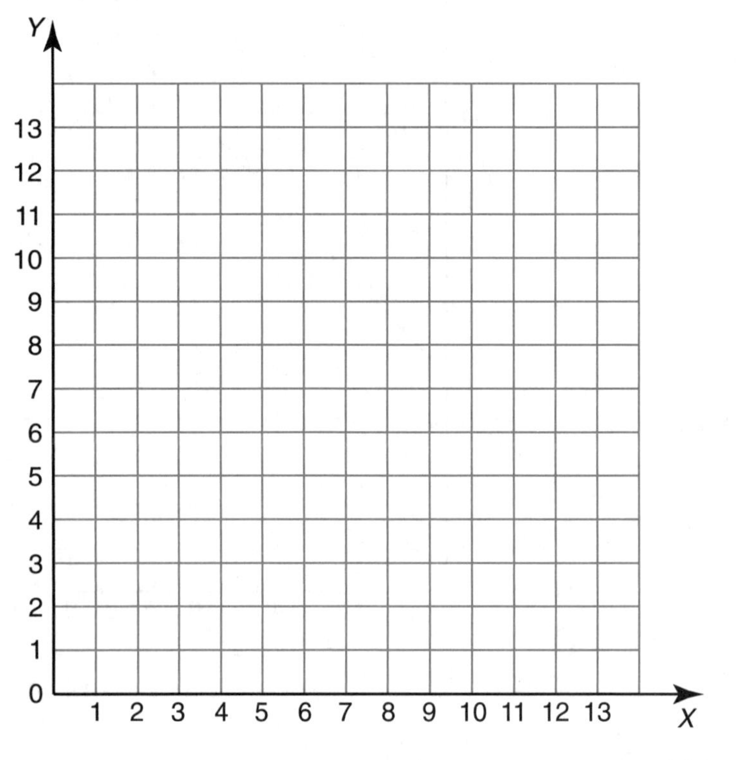

Answers on pages 189 to 191.

EXTENDED CONSTRUCTED-RESPONSE ITEMS

1. Each year on her birthday, Isabella's grandmother gave her $250.00 starting on her fifth birthday. She saved her money each year as shown in the table below.

Birthday	5	6	7	8	9	10
Money saved	$250	$500	$750	$1000	$1250	

- If Isabella's grandmother continues to give her $250 on each successive birthday, how much will Isabella have saved after year 12? Explain your answer.
- Isabella wants to buy a car when she becomes 17. The used car costs $5,000. Will Isabella have enough money to buy the car? Show your work and explain your answer.

Answers on page 191.

2. ■ Lani drew the pattern below. Draw the next three figures in the pattern and describe the rule.

$$\Delta\ \Delta\ \phi\ \Delta\ \phi\ \Delta\ \Delta\ \phi\ \Delta\ \phi\ \Delta\ \Delta$$

■ Sarah wrote the pattern of numbers below. What are the missing numbers? What is the rule for Sarah's pattern?

5, 10, 12, 24, 26, __, __, 108, 110

Answers on page 191.

3. ■ The scale below shows piggy banks and gold coins balanced. Write the number sentence that represents the model.

■ How many gold coins balance the piggy bank? Show all work and explain your answer.

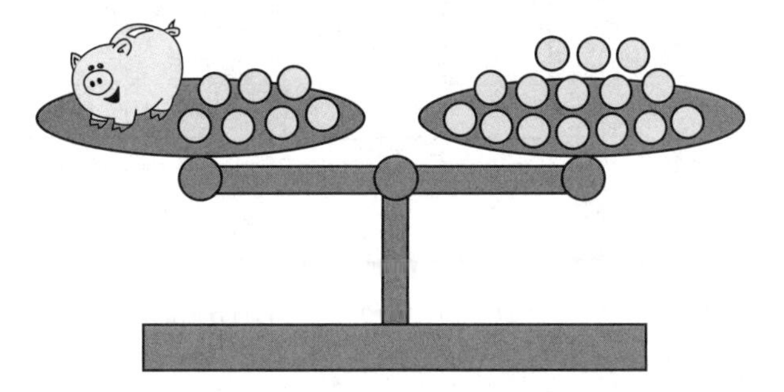

Answers on page 192.

4. ■ Complete the table with three ordered pairs that satisfy the equation $Y = 3X$. Show all work and explain your answer.

X	Y

- Use the coordinates in your table to graph the line that represents the equation $Y = 3X$. Show all work and explain your answer.

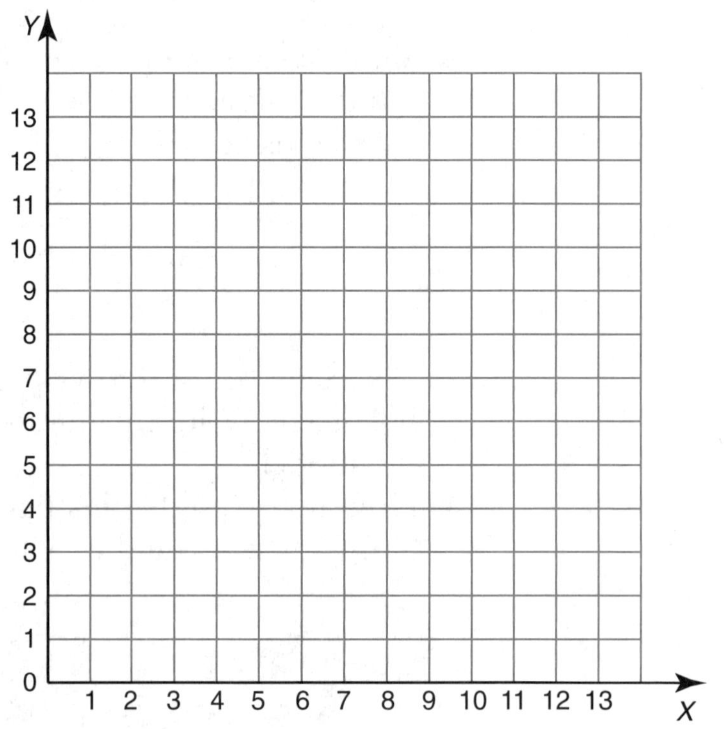

Answers on page 192.

Chapter 6

DATA ANALYSIS, PROBABILITY, AND DISCRETE MATHEMATICS

Data analysis, probability, and discrete mathematics are important interrelated areas of applied mathematics. Each provides students with powerful mathematical perspectives on everyday phenomena and with important examples of how mathematics is used in the modern world.

Data Analysis (or Statistics). In today's information-based world, you need to be able to read, understand, and interpret data in order to make informed decisions. You should be involved in collecting and organizing data and in presenting it using tables, charts, and graphs.

Probability. You need to understand the fundamental concepts of probability so that they can interpret weather forecasts, avoid unfair games of chance, and make informed decisions about medical treatments whose success rate is provided in terms of percentages. You should regularly be engaged in predicting and determining probabilities.

Discrete Mathematics—Systematic Listing and Counting. You should develop strategies for listing and counting.

Discrete Mathematics—Vertex-Edge Graphs and Algorithms. Vertex-edge graphs, consisting of dots (vertices) and lines joining them (edges), can be used to represent and solve problems based on real-world situations.

These topics provide students with insight into how mathematics is used by decision makers in our society, and with important tools for modeling a variety of real-world situations. Students will better understand and interpret the vast amounts of quantitative data that they are exposed to daily, and they will be able to judge the validity of data-supported arguments. (see *http://www.nj.gov/education/cccs/s4_math.pdf.*)

In this section you will learn how to interpret data using the **Three C's** which are Collecting Data, Comprehending Data, and Comparing Data. In **collecting data,** you will examine surveys. In **comprehending data,** you will learn how to understand the data using *mean, median, range, and mode.* Lastly, you will discover how to organize your data in many different ways in **comparing data.**

COLLECTING DATA

There are many ways of collecting data. Some of the most common ways of collecting data are through surveys, interviews, experiments, and research. In this section we are going to focus on surveys. To conduct a survey find *one* question to ask different people and then record their answers.

Example. Shayna and Dana conducted the following survey. "What is your favorite sport?" Below are the results of the survey.

Sport	Number Who Prefer Sport
Volleyball	2
Baseball	6
Basketball	4
Football	5
Tennis	1
Other	2

What fraction of the class prefers football?

First identify the number who chose football. **5**

Then determine the total number surveyed. 2 + 6 + 4 + 5 + 1 + 2 = **20**

$$\frac{\text{Fraction who prefer football} = \text{Number who prefer football}}{\text{Total number surveyed}} = \frac{5}{20} = \frac{1}{4}$$

COLLECTING DATA EXERCISES

Directions: Use the following information to answer questions 1 through 3.

A group of fifth grade students and a group of fourth grade students were asked this survey question, "What is your favorite type of music?" The following table shows the results of the survey.

What Is Your Favorite Type of Music?

	Rap	Hip Hop	R&B
Fifth Graders	2	7	12
Fourth Graders	1	14	6

1. What fraction of the fifth graders prefers Hip Hop?

A. $\frac{2}{3}$

B. $\frac{1}{3}$

C. $\frac{1}{7}$

D. $\frac{1}{2}$

2. What fraction of the fourth graders prefers Hip Hop?

A. $\frac{2}{3}$

B. $\frac{1}{3}$

C. $\frac{1}{7}$

D. $\frac{1}{2}$

3. What fraction of fourth and fifth graders prefers Hip Hop?

A. $\frac{2}{3}$

B. $\frac{1}{3}$

C. $\frac{1}{7}$

D. $\frac{1}{2}$

4. What fraction of the fifth graders prefers R&B?

A. $\frac{2}{7}$

B. $\frac{1}{7}$

C. $\frac{1}{3}$

D. $\frac{4}{7}$

5. What fraction of fourth graders prefers R&B?

A. $\frac{2}{7}$

B. $\frac{1}{7}$

C. $\frac{1}{3}$

D. $\frac{4}{7}$

Answers on page 192.

COMPREHENDING DATA

The range, mean, median, and mode are numbers that expresses certain sets of data.

RANGE

The **range** is the difference between the highest and the lowest numbers in a set of numbers. To find the range, subtract the lowest number from the highest.

Example 1. Six students conducted a Basketball Free Throw Experiment by shooting 10 free throws. Here is a list of the number of free throws made for each student.

8, 2, 4, 1, 6, 3

What is the range of free throws made?

Find the largest and the smallest number from the set of data.

Largest: 8   Smallest: 1

Subtract the smallest number from the largest number.

$$8 - 1 = 7$$

The range of data is 7.

MEDIAN

The **median** is the middle number when numbers are arranged in order. If there are two middle numbers, the median is the average of the two. If there are two numbers in the middle, add them and divide by two. The trick is "med" sounds like middle so you can remember that it's the middle number.

Example 2. Mr. Smith's students received the following scores on their math test.

84, 97, 68, 75, 93, 86, 72

Order the numbers from least to greatest.

68, 72, 75, 84, 86, 93, 97

Identify the middle number.

Since there are seven numbers, there is exactly one middle number. The median is 84.

Example 3. When there is an even amount of numbers, then there are two middle numbers. In this case, find the sum of the two middle numbers and divide by 2. One student took the math test the next day and the score was an 89. Mr. Smith added this score in with the other scores.

68, 72, 75, 84, 86, 89, 93, 97

Now there is an even number of scores and 2 middle numbers: 84 and 86

$$\text{Median of an even data set} = \frac{\text{Sum of the two middle numbers}}{2}$$

$$= \frac{(84+86)}{2} = \frac{170}{2} = 85$$

MEAN

The **mean** of a set of numbers is found by dividing the sum of the numbers by how many numbers there are. It is also known as the average.

Example 4. A list of Giana's wait times for the most popular rides at Rockin Rollercoaster Amusement Park in minutes follows:

$$32, 40, 4, 21, 13$$

What is the mean of Giana's minutes waiting on line at Rockin Rollercoaster Amusement Park?

Find the sum of all the numbers. Add to get the sum.

$$32 + 40 + 4 + 21 + 13 = 110$$

Count how many numbers there are. There are 5 numbers.

Divide the sum of number by how many numbers there are.

$$\text{Mean} = \frac{\text{Sum of numbers}}{\text{How many numbers}} = \frac{110}{5} = 22 \text{ minutes}$$

The mean of Giana's minutes on line is 22 minutes.

MODE

The **mode** is the number that occurs most often in a set of data.

Example 5. The daily high temperatures in Sewell, New Jersey, for the past week were 87, 86, 89, 98, 92, 89, and 92. What is the mode of these high temperatures? Find the temperature that occurs most often. There are two temperatures that occur two times so there are two modes: 89 and 92.

Example 6. What is the mode for this set of data: 1, 2, 3, 3, 4, and 5? The mode is 3.

Example 7. What is the mode for this set of data: 6, 17, 8, 19, 1, and 10? There is no mode, because no number occurs more often than any other number.

COMPREHENDING DATA EXERCISES

1. 8, 4, 9, 2, 5, 6, 3, 6, 2 Range _____ Median _____ Mean _____ Mode _____	2. 6, 9, 3, 1, 9, 8 Range _____ Median _____ Mean _____ Mode _____	3. 12, 1, 10, 4, 8, 6, 1, 2 Range _____ Median _____ Mean _____ Mode _____
4. What is the range of this data? 2, 7, 3, 15, 1, 8, 10 A. 8 B. 10 C. 14 D. 7	5. What is the mode of this data? 22, 11, 33, 22, 12, 5 A. 11 B. 28 C. None D. 22	6. What is the median of this data? 14, 21, 19, 16, 20 A. 18 B. 19 C. 20 D. 6

Timmy kept a record of how many hours he watched TV in one week.

Day of the Week	Hours Timmy Watched TV
Sunday	3
Monday	2
Tuesday	1
Wednesday	2
Thursday	6
Friday	4
Saturday	3

7. What is the mode of the above data?
   - A. 2
   - B. 6
   - C. 4
   - D. 2, 3

8. What is the mean of the above data set?
   - A. 6
   - B. 5
   - C. 3
   - D. 4

9. What is the median of the above data set?
   - A. 6
   - B. 5
   - C. 3
   - D. 4

10. What is the range of the above data set?

A. 6

B. 5

C. 3

D. 4

Answers on pages 192 to 193.

COMPARING DATA—BAR GRAPHS

Bar graphs can be used to compare data and represent the trend of data. The **scale** of a graph shows the number increments that the x and y axes increase. The **labels** on the x and y axes of the graph tells you what each part of the graph represents. It is important for the graph to have a title, too.

Example. Below is six months of data that shows the number of vanilla and strawberry ice creams sold in the Monongahela Middle School cafeteria during a six month period.

Ice Cream Flavor	January	February	March	April	May	June
Vanilla	35	25	41	53	70	10
Strawberry	14	15	28	17	32	5

Below is a bar graph representing the same data in the table.

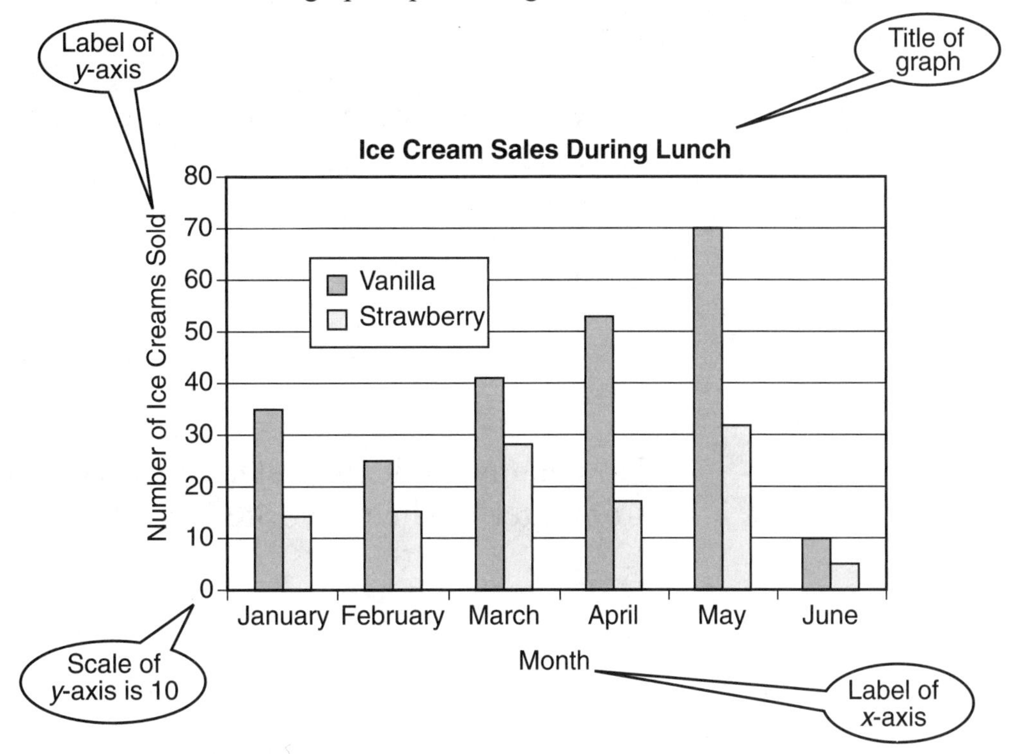

The bar graph makes it easier to see that more students purchased vanilla ice cream than strawberry in the 6-month period. The bar graph also shows that more ice cream was purchased throughout the spring months.

COMPARING DATA—BAR GRAPHS EXERCISES

Below is a bar graph of data from a survey done on the students at Anytown Elementary School.

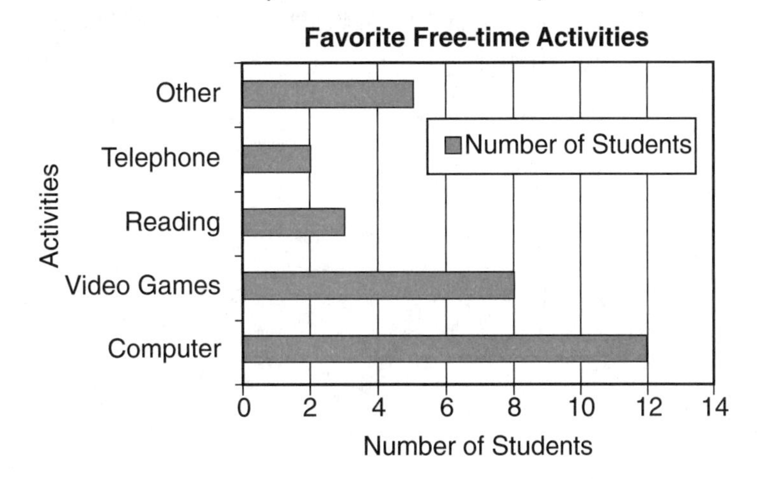

Directions: Use the above bar graph to answer the following questions.

1. How many students prefer reading as their favorite activity?

   A. 2.5

   B. 3

   C. 2

   D. 5

2. How many students prefer Video Games and Computer as their favorite activity?

   A. 20

   B. 12

   C. 8

   D. 4

3. Which activity did the students prefer the most?

   A. Video games

   B. Computer

   C. Reading

   D. Other

4. How many more students preferred video games over reading?

   A. 5.5

   B. 2.5

   C. 5

   D. 8

5. How many students chose Other as their favorite free-time activity?

   A. 4.5

   B. 5.5

   C. 6

   D. 5

6. What is the scale of the x-axis?

   A. 7

   B. 14

   C. 2

   D. None of the above

7. How many more students chose Computers over Telephone? _____

8. What activity is the least favorite for this group of students? _____

9. How many students were surveyed in all? _____

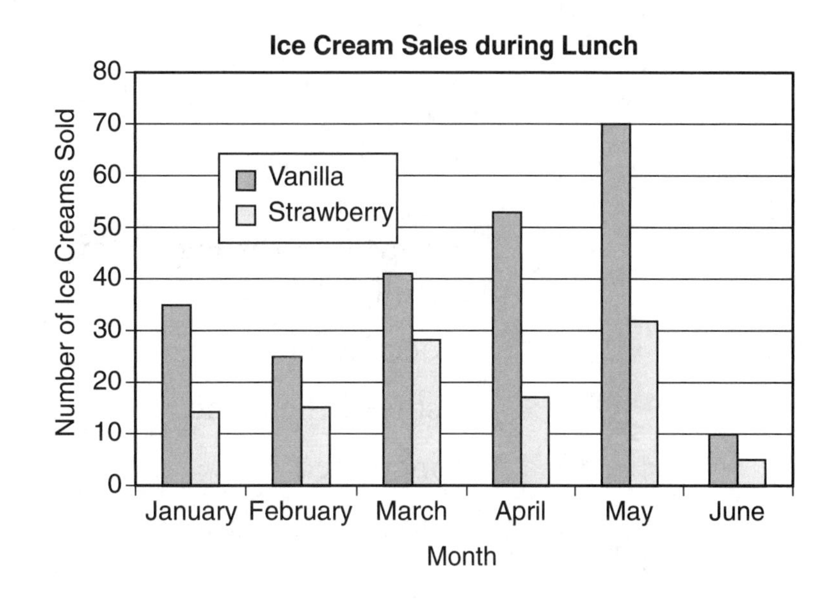

Directions: Use the above bar graph to answer questions 10 through 14.

10. Approximately how many vanilla ice creams were sold from January 1 to March 31?

    A. 40

    B. 35

    C. 100

    D. 75

11. In what month did the cafeteria sell the most ice cream?

    A. April

    B. May

    C. January

    D. February

12. How many more vanilla ice creams were sold than strawberry ice creams in February?

    A. 5

    B. 15

    C. 12

    D. 10

13. From March through May, the sale of ice cream generally _____.

    A. decreased

    B. stayed the same

    C. increased

    D. none of the above

14. In what month did the cafeteria sell 15 ice creams?

A. June

B. February

C. January

D. none of the above

Answers on pages 193 to 194.

COMPARING DATA—LINE GRAPHS

A **line graph** shows information like trends or changes in data over a period of time. A **trend** is a pattern that the data follows over time.

Example. Below is a line graph of how much money Mandy's Handy Candy Shop earns in one week.

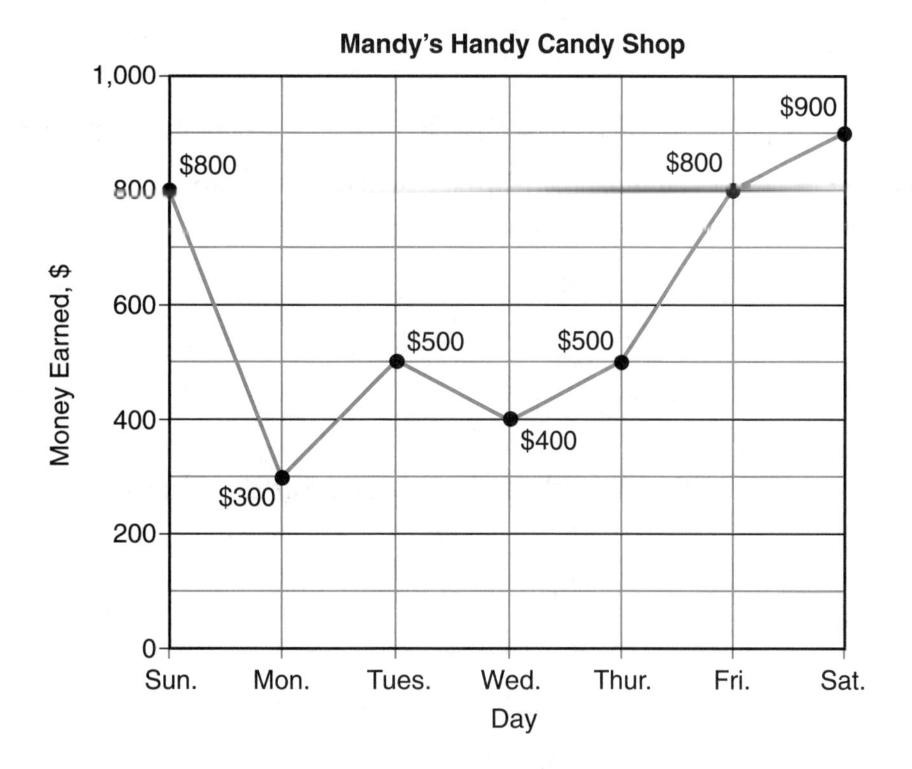

COMPARING DATA—LINE GRAPHS EXERCISES

Directions: Use the line graph in the Example to answer these questions.

1. What is the scale of the y-axis?
   A. 50
   B. 100
   C. 200
   D. 500

2. What is the mode of the data?
   A. $50
   B. $100
   C. $200
   D. $500

3. What is the difference between the amount of money earned on Monday and Thursday?
   A. $50
   B. $100
   C. $200
   D. $500

4. On what day does Mandy's Handy Candy Shop earn more money?
   A. Friday
   B. Saturday
   C. Sunday
   D. Monday

5. On what day does Mandy's Handy Candy Shop earn $300?
   A. Friday
   B. Saturday
   C. Sunday
   D. Monday

6. Is there a trend shown in the graph? If so, what is it?

Answers on page 194.

COMPARING DATA—CIRCLE GRAPHS

Circle graphs, also known as pie graphs, are divided into sections. Each section uses a percent or a fraction to show a piece of the whole. The total sections of a graph add up to 100% or 1 if you use fractions.

COMPARING DATA—CIRCLE GRAPHS EXERCISES

Mr. Han kept track of all the activities his students do over the weekend. He made the following circle graph to display his data.

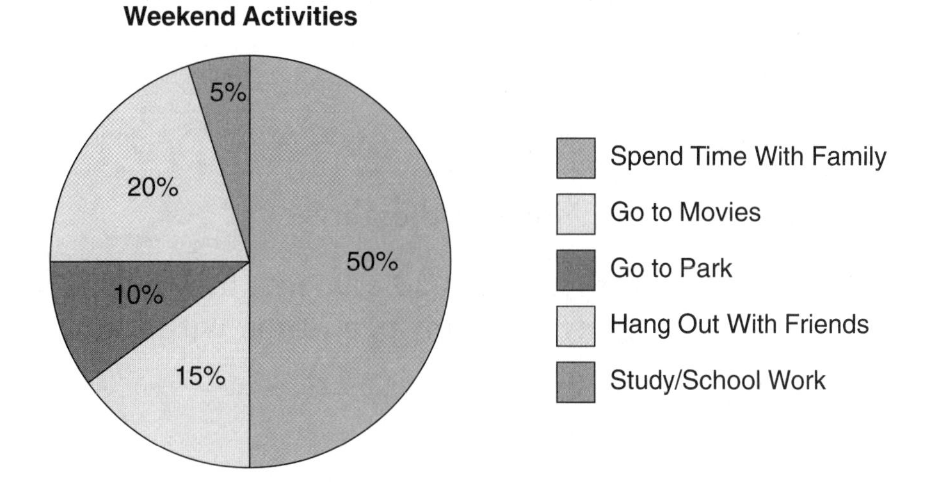

Directions: Use the circle graph from the example above to answer questions 1 through 5.

1. How do most of the students spend their time on the weekend?
   A. Study/school work
   B. Hang out with friends
   C. Spend time with family
   D. Go to the movies

2. What activity did the least number of students do?
   A. Study/school work
   B. Hang out with friends
   C. Spend time with family
   D. Go to the movies

3. What activity did 20% of students do over the weekend?
   A. Study/school work
   B. Hang out with friends
   C. Spend time with family
   D. Go to the movies

4. What is the percent difference between hanging out with friends and study/school work?
   A. 20%
   B. 5%
   C. 15%
   D. None of the above

5. If there were 20 students in Mr. Han's class, how many of them spend time with family?
   A. 20
   B. 10
   C. 15
   D. None of the above

Answers on page 194.

INTUITIVE PROBABILITY

The probability of an event can be expressed in terms of how likely or unlikely it is to occur. If an event is impossible, it will never occur and the probability is 0%. An event that is certain to occur is 100% likely to occur.

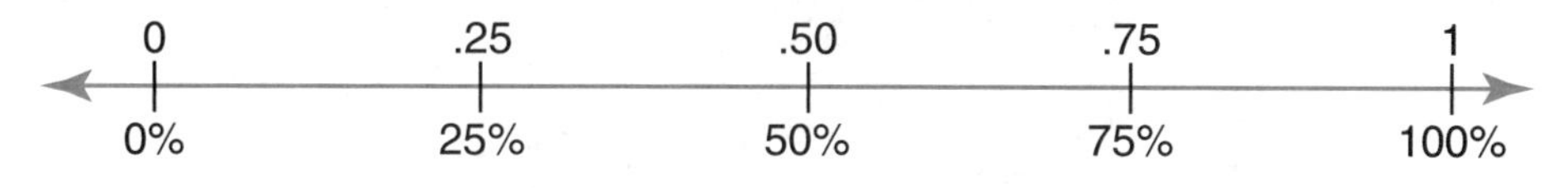

Example 1. Take a spin. What is the probability that you will land on red?

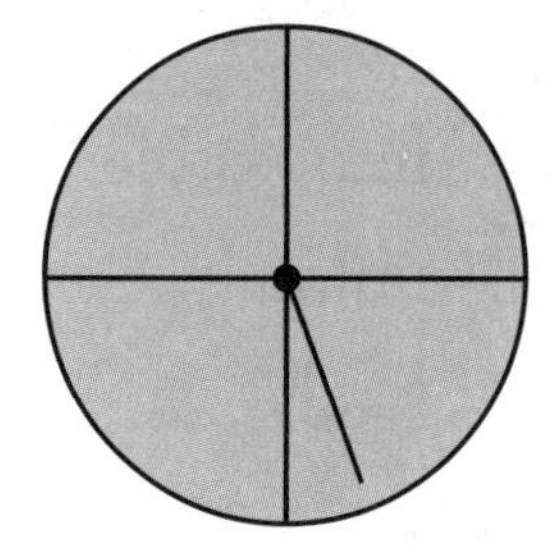

There are 4 total outcomes and all four of the outcomes are red.

$$\frac{4}{4} = 1$$

Example 2. There are five Jolly Ranchers and five Starbursts in the bag.

If you picked out six pieces of candy with blindfolds on, what is the probability of blindly picking out six Jolly Ranchers?

The probability is 0 because there are only five Jolly Ranchers. It is impossible to draw six Jolly Ranchers if there are only five in the bag.

Answers on pages 196 to 197.

Example 3. Express the probability of the spinner landing on the number 2 in terms of one of the following:

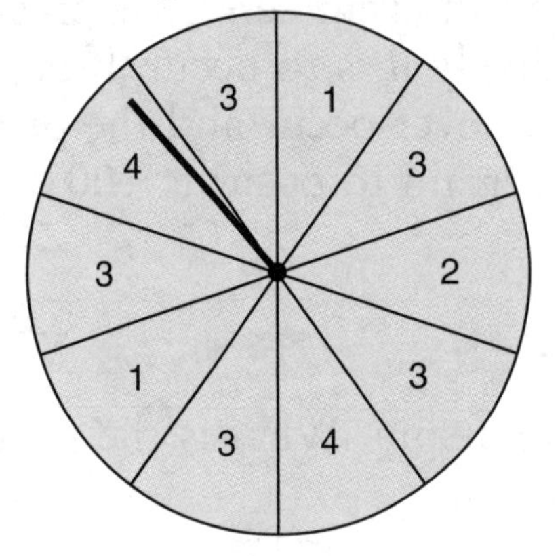

A. impossible
B. unlikely
C. likely
D. certain

The probability of spinning the number 2 is unlikely (B). There are a total of 10 possible outcomes.

The number 2 is 1 of the 10 possible outcomes. The probability is $\frac{1}{10}$.

INTUITIVE PROBABILITY EXERCISES

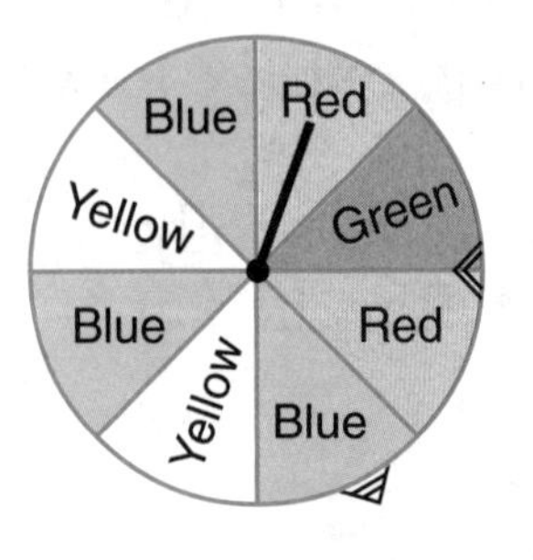

1. What is the probability of landing on green?
   A. Likely
   B. Equally likely
   C. Less likely

2. What is the probability of landing on 1?
   A. Likely
   B. Equally likely
   C. Less likely

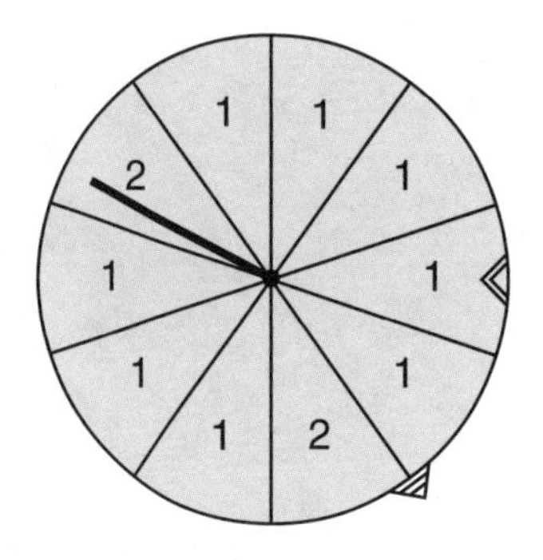

3. What color lollipop has the greater probability of being chosen? _____

4. What color lollipops have the same probability of being chosen? _____

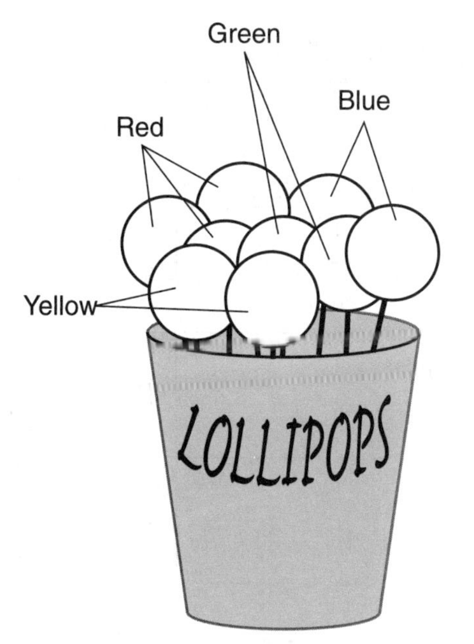

Answers on pages 194 to 195.

EXPERIMENTAL PROBABILITY

Experimental probability is the ratio of the number of favorable outcomes to the total number of outcomes in an experiment.

$$\textbf{Probability (P)} = \frac{\textbf{Number of favorable outcomes in an experiment}}{\textbf{Total number of outcomes in an experiment}}$$

Example

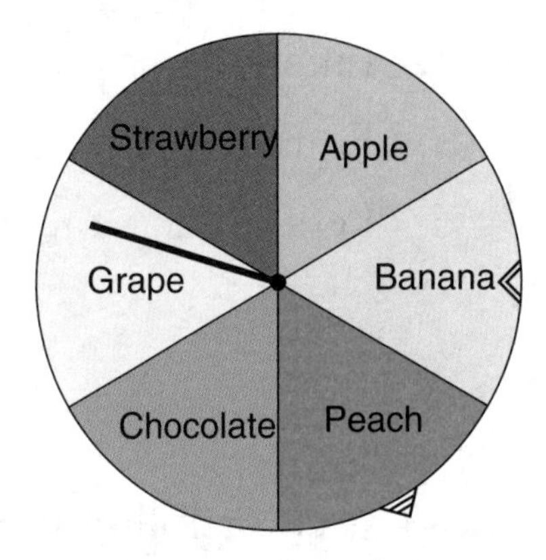

The total number of outcomes (spins) is 20. Out of 20 spins, the spinner landed on Strawberry 4 times.

Bubble Gum Monkey Candy	Frequency of Times Spinner Lands on Flavor	Experimental Probability of Landing on Flavor
Strawberry	4	$P(s) = \frac{4}{20}$
Apple	1	$P(A) = \frac{1}{20}$
Banana	7	$P(B) = \frac{7}{20}$
Peach	2	$P(P) = \frac{2}{20} = \frac{1}{10}$
Chocolate	4	$P(C) = \frac{4}{20} = \frac{1}{5}$
Grape	2	$P(G) = \frac{2}{20} = \frac{1}{10}$

EXPERIMENTAL PROBABILITY EXERCISES

1. The total number of rolls on a six-sided die is 15. Fill in the table with the probability of rolling the following.

Favorable Outcomes of a Six-Sided Die	Frequency of Times Number Is Rolled	Experimental Probability Rolling a Number
1	2	P (1) =
2	1	P (2) =
3	0	P (3) =
4	4	P (4) =
5	3	P (5) =
6	5	P (6) =

2. To determine the experimental probability of flipping heads or tails up, fill in the following table.

Favorable Outcomes	Frequency of Outcome	Experimental Probability
Head	8	P(H) =
Tail	4	P(T) =

Answers on page 195.

THEORETICAL PROBABILITY

Theoretical probability is the ratio of the number of favorable outcomes to the total number of possible outcomes.

$$\text{Probability (P)} = \frac{\textbf{Number of favorable outcomes}}{\textbf{Total number of possible outcomes}}$$

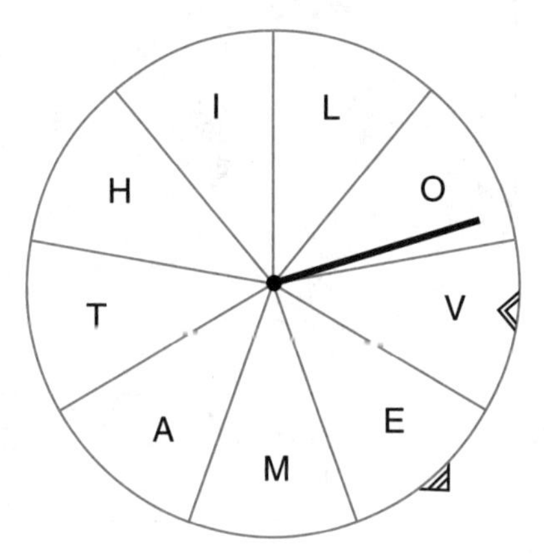

Example 1. What is the theoretical probability of the spinner landing on a vowel?

There are four vowels: I, O, E, A.
There are nine possible outcomes: I,L,O,V,E,M,A,T,H.

$$\text{P(vowel)} = \frac{4}{9}$$

What is the theoretical probability of the spinner landing on a consonant?

There are five consonants: L, V, M, T, H.

$$\text{P(consonant)} = \frac{5}{9}$$

Example 2. What is the theoretical probability of rolling a factor of 6?

This net shows all six sides of the same die.
There are four factors of 6: 1, 2, 3, 6.
There are six possible outcomes: 1, 2, 3, 4, 5, 6.

$$P(\text{factor of } 6) = \frac{4}{6} = \frac{2}{3}$$

Example 3. What is the theoretical probability of drawing a dark marble, a white marble, and a checkered marble? There are 2 dark marbles, 3 checkered marbles, and 5 white marbles.

There 10 possible outcomes: 10 total marbles

$$P(\text{dark marble}) = \frac{2}{10} = \frac{1}{5} \qquad P(\text{checkered marble}) = \frac{3}{10}$$

$$P(\text{white marble}) = \frac{5}{10} = \frac{1}{2}$$

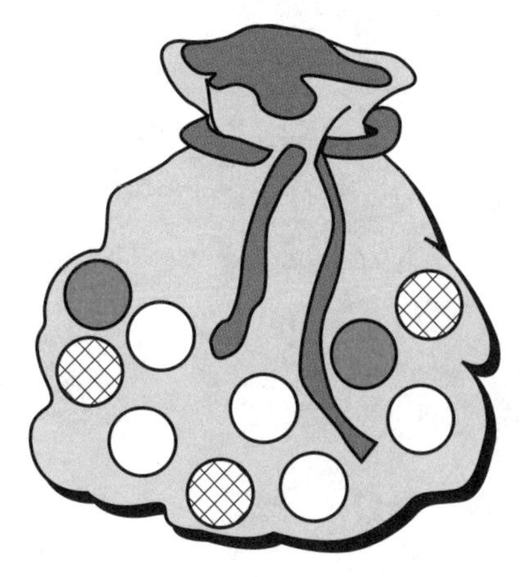

THEORETICAL PROBABILITY EXERCISES

1. If you close your eyes, what is the probability of choosing a boy's name?

2. If you close your eyes, what is the probability of choosing a girl's name?

3. If you close your eyes, what is the probability of choosing a name that begins with a vowel?

Directions: Use the net view of the die to answer the following questions.

4. What is the probability of rolling an even number?

A. $\frac{2}{3}$

B. $\frac{1}{2}$

C. $\frac{1}{6}$

D. $\frac{1}{3}$

5. What is the probability of rolling a multiple of three?

A. $\frac{2}{3}$

B. $\frac{1}{2}$

C. $\frac{1}{6}$

D. $\frac{1}{3}$

6. What is the probability of rolling a number greater than 5?

A. $\frac{2}{3}$

B. $\frac{1}{2}$

C. $\frac{1}{6}$

D. $\frac{1}{3}$

Directions: Use the bag of marbles to answer the following.

7. What is the probability of drawing a checkered marble without looking?

A. $\frac{2}{5}$

B. $\frac{1}{3}$

C. $\frac{1}{15}$

D. $\frac{1}{5}$

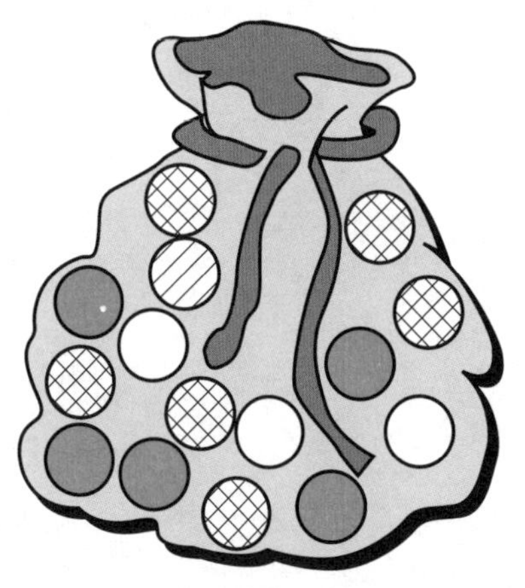

8. What is the probability of drawing a white marble without looking?

A. $\frac{2}{5}$

B. $\frac{1}{3}$

C. $\frac{1}{15}$

D. $\frac{1}{5}$

9. What is the probability of drawing a striped marble without looking?

A. $\frac{2}{5}$

B. $\frac{1}{3}$

C. $\frac{1}{15}$

D. $\frac{1}{5}$

10. What is the probability of drawing a dark marble without looking?

A. $\frac{2}{5}$

B. $\frac{1}{3}$

C. $\frac{1}{15}$

D. $\frac{1}{5}$

Answers on pages 196 to 197.

MULTIPLICATION PRINCIPLE OF COUNTING

The **multiplication principle** involves the use of multiplication as a strategy to determine all possible outcomes or combinations when calculating probabilities. If you can make the first choice in ***m*** ways and the second choice in ***n*** ways, the total possible outcomes can be determined by multiplying ***m*** ways × ***n*** ways.

Example 1. How many possible combinations are there to choose an outfit if there are 4 pairs of pants, 2 shirts, and 2 pairs of sneakers. The tree diagram below shows all of the possible outcomes.

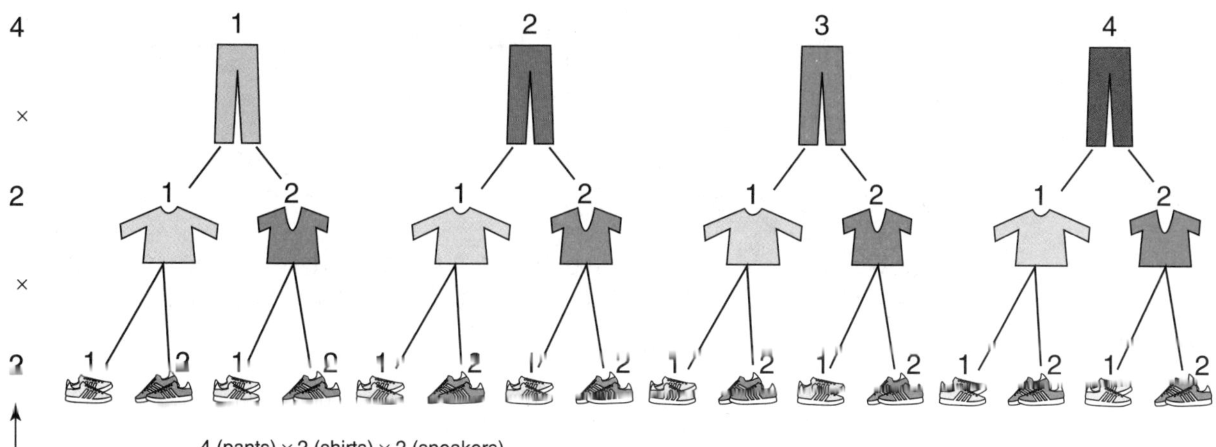

Example 2. At Cheeseburger Madness, they have 5 different kinds of sandwiches, 3 different kinds of soda, and 2 different kinds of dessert. How many different combinations of a sandwich, soda, and dessert are there?

Example 3. Juan, Carmen, and Jenny are in a race. What is the probability that Jenny will come in first place?

First, we have to find the possible outcomes. The list below shows all of the possible outcomes of the race.

1st, 2nd, 3rd Place

Juan, Carmen, Jenny

Juan, Jenny, Carmen

Carmen, Juan, Jenny

Carmen, Jenny, Juan

Jenny, Juan, Carmen

Jenny, Carmen, Juan

There are 6 possible outcomes.

Below is another way to find the possible outcomes of the race.

How many different students can possibly come in first place? 3

How many different students can possibly come in second place if one has already won first place? 2

How many different students can possibly come in third place if one student has already won first and another student has won second? 1

$$\begin{array}{l}\mathbf{1^{st}, 2^{nd}, 3^{rd}\ Place}\\ \downarrow\ \ \downarrow\ \ \ \downarrow\\ 3\times 2\times 1 = 6 \text{ possible outcomes}\end{array}$$

According to the list above, Jenny could possibly come in first 2 out of the 6 possible outcomes.

$$P(1^{st}\text{ Place}) = \frac{2}{6} = \frac{1}{3}$$

Example 4. Jalen has a six-sided die numbered 1 through 6. He is playing Fortune Five. In order to win, he has to roll the die two times and have the sum of his two rolls add up to five. How many ways can his rolls add up to five?

$$1+4, 2+3, 3+2, 4+1$$

There are 4 ways to win the Fortune Five game. Rolling a five or six would be losers.

MULTIPLICATION PRINCIPLE OF COUNTING EXERCISES

1. So Good Ice Cream Parlor offers 10 different ice cream flavors, 2 different kinds of cones, and 4 different toppings. How many days can Shawn buy ice cream without having the same ice cream combination twice?

   A. 20 days

   B. 40 days

   C. 16 days

   D. 80 days

2. During breakfast, the school cafeteria offers chocolate milk, strawberry milk, regular milk, or orange juice. Students can also choose a muffin, pastry, or bagel. According to the tree diagram, how many choices does each student have?

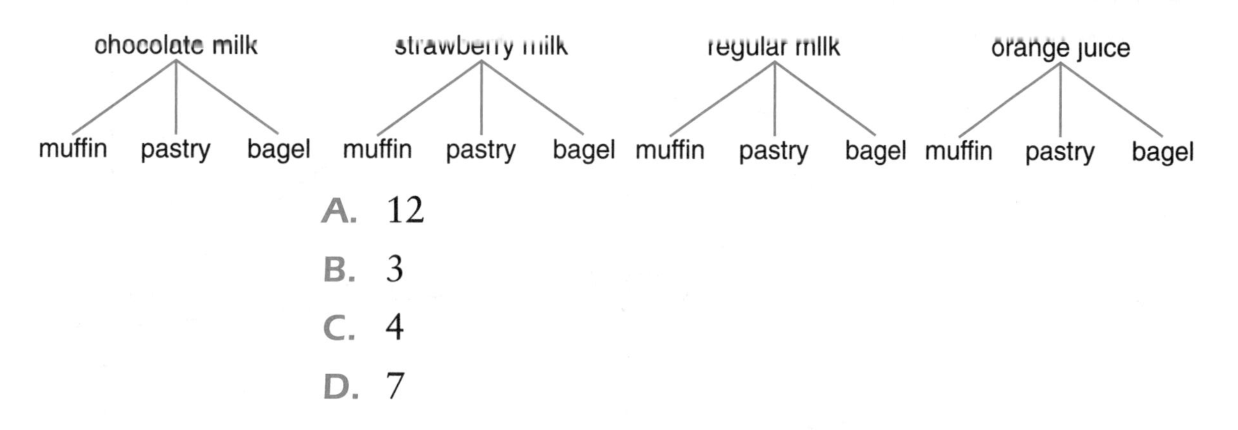

   A. 12

   B. 3

   C. 4

   D. 7

3. Find the answer for questions using the multiplication principle of counting.

4. Jim, Veronica, and Mia have received the most tickets for completion of homework. Their names will be put in a hat and chosen at random to be awarded first place, second place, and third place prizes. What is the

probability that Mia will not get the first place prize? (Hint: Make a list of combinations.)

5. Rebecca has a six-sided die numbered 1 through 6. She is playing Lucky Seven. In order to win, she has to roll the die two times and have the sum of her two rolls add up to seven. How many ways can her rolls add up to Lucky Seven?

Answers on pages 197 to 198.

VERTEX-EDGE GRAPHS

A **vertex-edge graph** is a graph consisting of a set of points (called vertices) joined by two or more edges (sides).

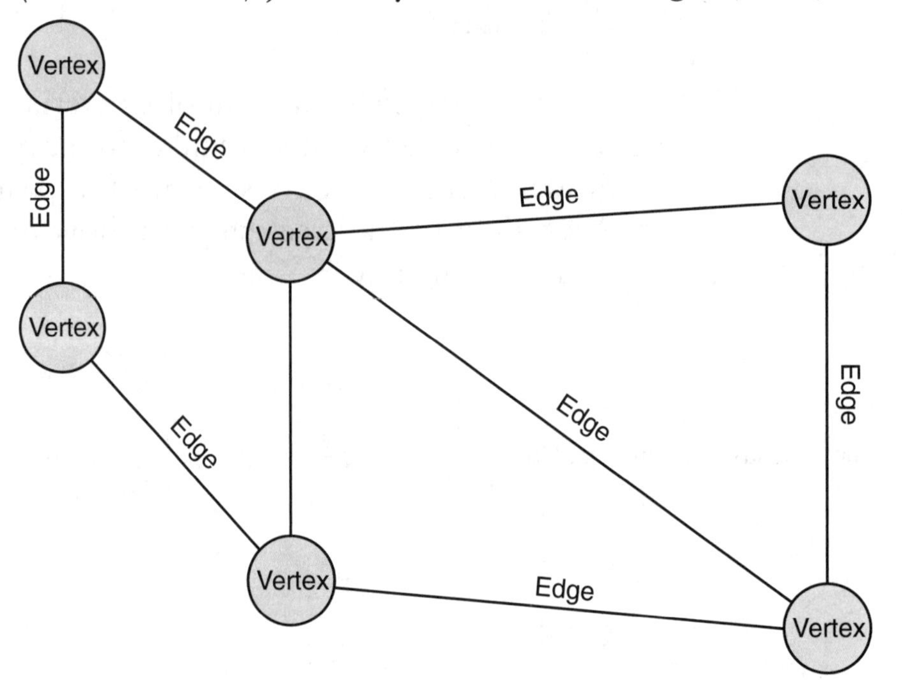

Example

In the following diagram, the distances between the vertices are shown. The shortest path from home to school is the one that transverses from home to playground to school, because it only travels along 2 edges and 1 vertex.

All other paths from home to school are longer because they are more than 2 edges and 1 vertex.

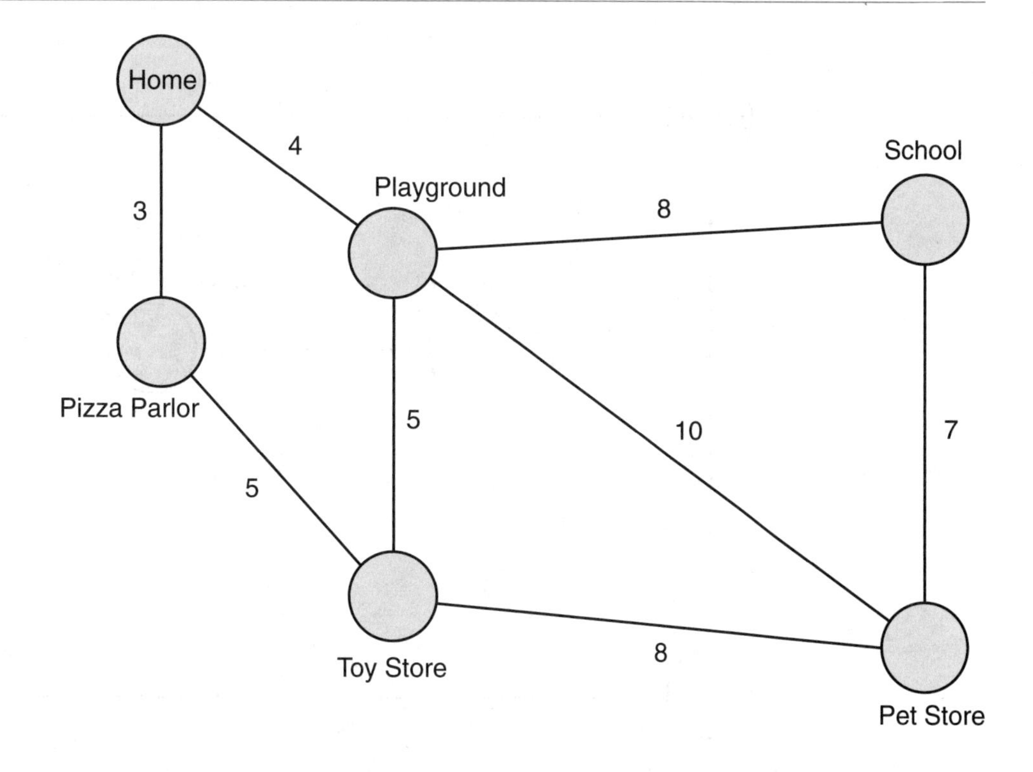

VERTEX-EDGE GRAPHS EXERCISES

1. Julian wants to rush home from the pet store to give his new puppy a toy ball. What is the distance in units of the shortest route from the pet store to home? Indicate your route by naming each vertex in order.

2. Julian has saved enough money to buy his puppy a toy bone. This time he wants to take his puppy to the pet store with him. He wants to take the longest route to the pet store to show off his new puppy to his friends. Of the routes below what is the longer route from home to the pet store?

   A. Home–Pizza Parlor–Toy Store–Pet Store

   B. Home–Playground–Toy Store–Pet Store

   C. Home–Playground–School–Pet Store

   D. Home–Playground–Pet Store

3. In the diagram above, what is the distance from school to the pizza parlor?

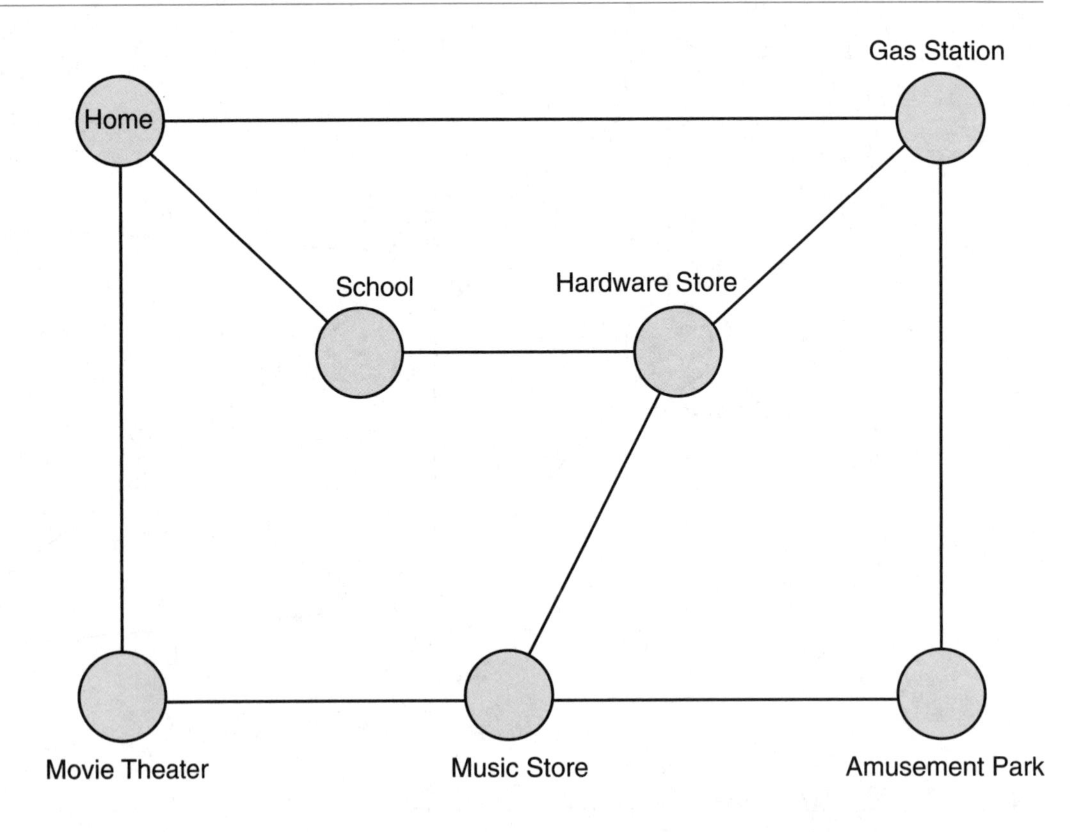

4. How many vertices are in the diagram above?
   A. 6
   B. 7
   C. 8
   D. None of the above

5. How many edges are in the diagram above?
   A. 9
   B. 8
   C. 7
   D. None of the above

Answers on page 198.

EXTENDED CONSTRUCTED-RESPONSE ITEMS

1. The circle graph below shows the breakdown of the most recent video game console owned by fifth grade students at Anytown Elementary School.

   If 50 fifth graders were surveyed, how many more fifth graders said that Video System A was their most recent video system than those that said Video System B was their most recent? Show all work and explain your answer.

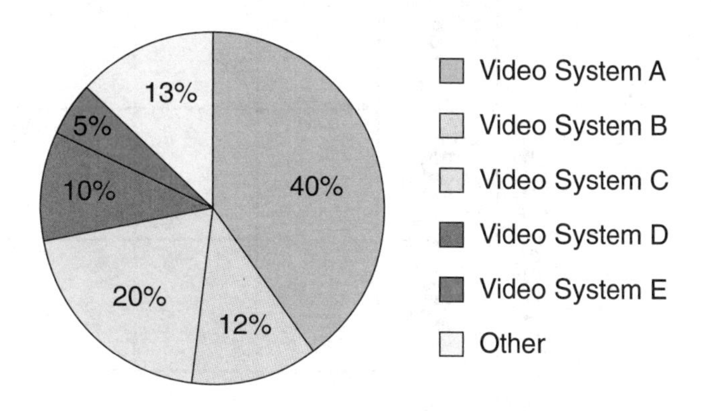

Answers on page 199.

2. A game consists of rolling two number cubes and adding the numbers on the two sides facing up. You win a prize if you roll a prime number!

   Fill in the chart with the sum of all possible outcomes of the two number cubes.

   How many of the outcomes from the chart below are prime numbers?

   What is the probability that you will win a prize when you roll two number cubes?

   Is this a fair game? Explain.

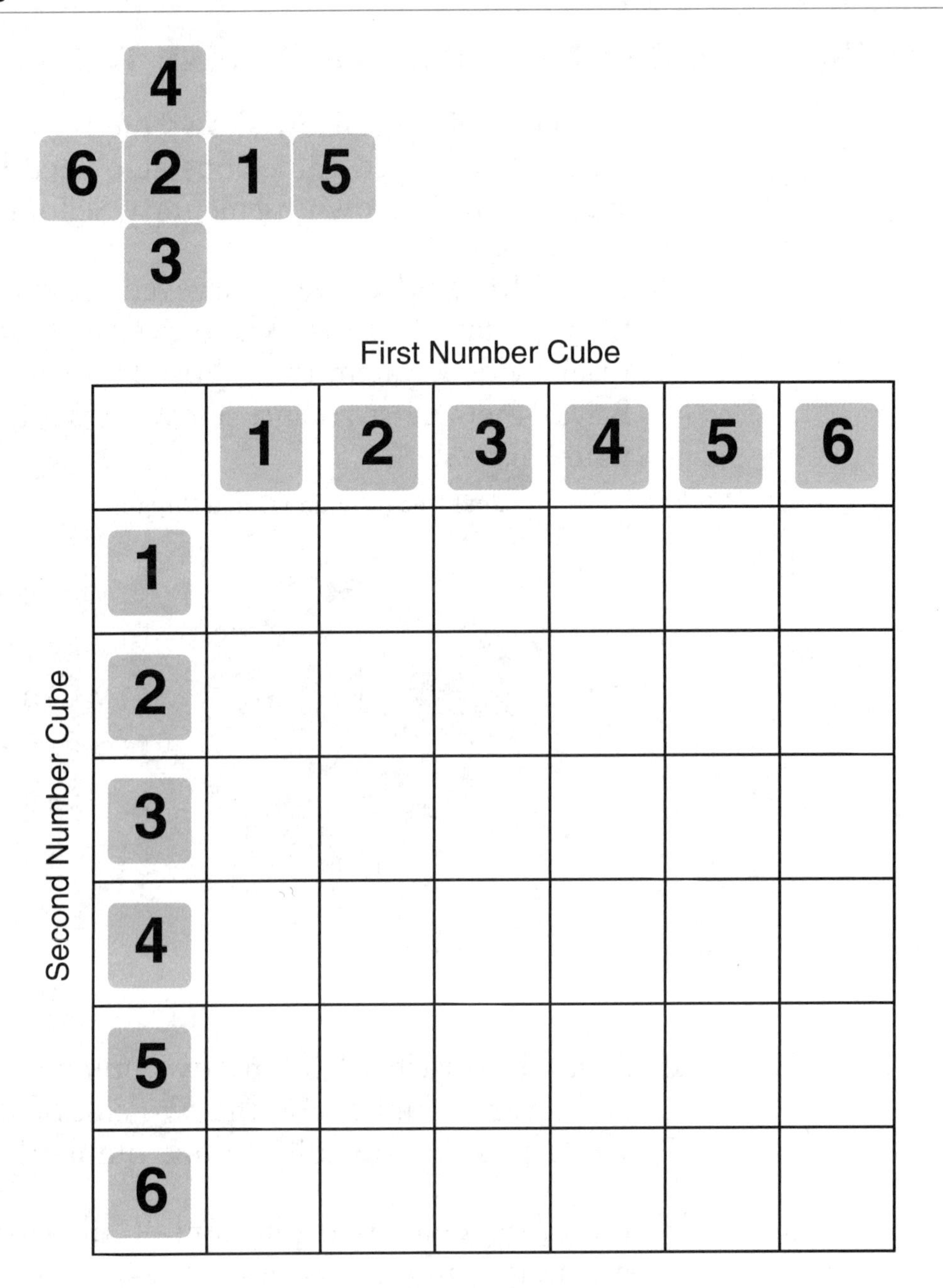

Answers on page 200.

3. Mrs. Jackson asked each student in her class to write down their favorite carnival/amusement park ride/attraction. Below is what each student wrote down.

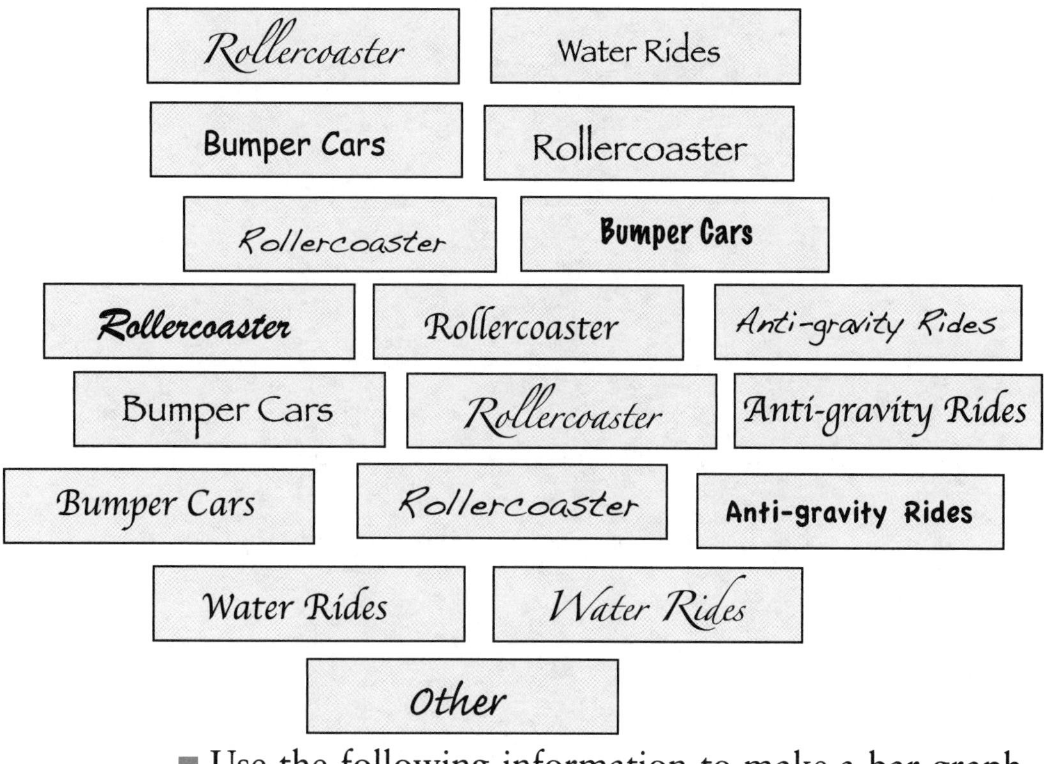

- Use the following information to make a bar graph. Choose a title, labels, and scale.
- Using the bar graph make a comparison between any two sets of data.

ANSWERS TO EXERCISES

CHAPTER 3

ANSWERS TO PLACE VALUE EXERCISES

1. 4,691,131
2. 912,433
3. 21,037
4. 462
5. 390,684
6. 74,507
7. 152,080
8. 16,334
9. 760,948
10. 907,469
11. thousands
12. ones
13. ten millions
14. tens

15. ten thousands

16. ten thousands

17. C

18. D

19. D

20. B

ANSWERS TO ESTIMATION EXERCISES

1. C, $30{,}000 + 45{,}000 = 75{,}000$

2. B, $40 \times 100 = 4{,}000$

3. D, $(6{,}000 + 8{,}000 + 4{,}000) = 18{,}000$

4. C, $40{,}000 - 30{,}000 = 10{,}000$

ANSWERS TO NUMERICAL OPERATIONS EXERCISES

1. You should multiply the 4 by $25 = 100 \times 9 = 900$

2. $900 - 200 = 700 - 1 = 699$

3. 2,106

4. 109

5. $(128 \times 2)/8 = 32$ ounces each

6. $\$780/6 = \$130$

7. $156/6 = 26$ packages

ANSWERS TO RATIONAL NUMBER EXERCISES

1. D, $\frac{1}{5} < \frac{1}{4} < \frac{1}{3} < \frac{1}{2}$

2. C, $\frac{3}{4} = 0.75$

3. B, $\frac{5}{6} = 0.8333 > 0.75$

4. C, 9.74

5. A, 12 – 9.5 = 2.5 gallons

6. 2 circles must be shaded $\frac{5}{10} = \frac{2}{4} = \frac{1}{2}$

7. $\frac{1}{2} + \frac{3}{8} = \frac{7}{8}$

8. D, \$5.00 – (2 × 75¢ + 50¢) = \$3.00

9. C, 1.75 × 2 = 3.5 miles

10. A, 0.5, $\frac{3}{5} = 0.6$, 1, 1.01

11. C, 3.1 – 1.95 = 1.15 miles

12. A, 2.05 + 0.01 = 2.06

13. 3 more students, $\frac{2}{3}$ of 36 = 24 voted for math, but we need $\frac{3}{4}$ of 36 = 27

14. 75% or 0.75 or $\frac{3}{4}$ of a gallon—all of these forms are acceptable.

15. Answers will vary. Possibilities include 0.21, 0.22, 0.23, 0.24, $\frac{21}{100}$.

ANSWERS TO NUMBER THEORY EXERCISES

1. A, 7 numbers: 10, 15, 20, 25, 30, 35, and 40
2. D, $90 = 2 \times 3 \times 3 \times 5$
3. A, 36 has prime factorization $2 \times 2 \times 3 \times 3$
4. C, $72/8 = 9$
5. B, a prime number has exactly two factors
6. B, 4, 8, 12, and 16 are all divisible by 2
7. C, 12 is the LCM of 4 and 6

EXTENDED CONSTRUCTED-RESPONSE ITEMS FOR CHAPTER 3

Question 1 You will receive full credit if:

You plot and label both points accurately, name a decimal and a fraction between $\frac{3}{5}$ and 0.8, and clearly explain why the chosen numbers meet the requirement.

Question 2 You will receive full credit if:

You answer all parts correctly including a clear explanation: 33 balls had a triangle, 8 balls had both a triangle and square, the 24th ball has all shapes painted on it because 24 is a multiple of 3, 4, and 6.

Question 3 You will receive full credit if:

You answer all parts correctly stating that 67 is prime but 57 is not prime, including a clear explanation.

CHAPTER 4

ANSWERS FOR TYPES OF ANGLES EXERCISES

1. Angle *MNO* or angle *ONM*
2. Point *N*
3. Ray *NM* and ray *NO*
4. Acute angle

ANSWERS FOR PARALLEL LINES EXERCISES

1. 3 pairs of parallel lines
2. 1 pair of parallel lines
3. 4 pairs of parallel lines

ANSWERS FOR PERPENDICULAR LINES EXERCISES

1. 1 pair of perpendicular lines
2. 4 pairs of perpendicular lines
3. 2 pairs of perpendicular lines

ANSWERS FOR INTERSECTING LINES EXERCISES

1. C. $\overleftrightarrow{AB}$ and $\overleftrightarrow{CD}$
2. $\overleftrightarrow{AB}$ and $\overleftrightarrow{CD}$

3. $\overleftrightarrow{AB}$ and $\overleftrightarrow{EH}$, $\overleftrightarrow{CD}$ and $\overleftrightarrow{EH}$, $\overleftrightarrow{AF}$ and $\overleftrightarrow{EF}$

4. $\overline{XY} \perp \overline{YZ}$, $\overline{WX} \perp \overline{WZ}$, $\overline{WX} \perp \overline{XY}$, $\overline{YZ} \perp \overline{WZ}$

ANSWERS FOR TRIANGLES EXERCISES

1. C. Equilateral Triangle, all sides are equal
2. B. Isosceles Triangle, two sides are equal
3. A. Right Triangle, it has one right angle
4. B. Scalene Triangle, all sides have different lengths
5. D. Scalene Triangle, all sides have different lengths

ANSWERS FOR QUADRILATERALS EXERCISES

1. B. Equilateral, all sides have the same measures.
2. A. Trapezoid, it has only one pair of sides parallel.
3. C. All squares have the properties of rectangles.
4. C. Isosceles trapezoid, opposite sides all equal.

ANSWERS FOR REGULAR POLYGONS EXERCISES

1. C, equilateral
2. B, equiangular
3. D, regular
4. D, pentagon
5. B, octagon
6. B, pentagons (like home plate in baseball) do not have to have all angles equal.

7. **D,** squares, by definition, always have equal angles and equal sides.

8. **D,** nonagon

9. **C,** pentagons have five sides.

10. **A,** right triangle, it has a 90 degree angle.

11. **A,** (12, 4)

ANSWERS FOR CIRCLES EXERCISES

1. *WX*

2. *XY*

3. *YZ*

4. *WY*

5. If the diameter is 8 units, the radius is 4 units.

6. If the radius is 7 units, the diameter is 14 units.

ANSWERS FOR CONGRUENT FIGURES EXERCISES

1. Triangles 1 and 4 are congruent (both are right triangles and exact copies).

2. C

ANSWERS FOR SIMILARITY EXERCISES

1. *X* and *Z*

2. *A* and *C*

3. Quadrilateral *LMNO* and quadrilateral *QRST*

ANSWER TO SIMILARITY AND SIDE LENGTH EXERCISE

1. **B,** the ratio between the two triangles is 1 : 3.

ANSWERS FOR SYMMETRY EXERCISES

1.

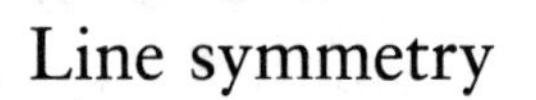

Line symmetry

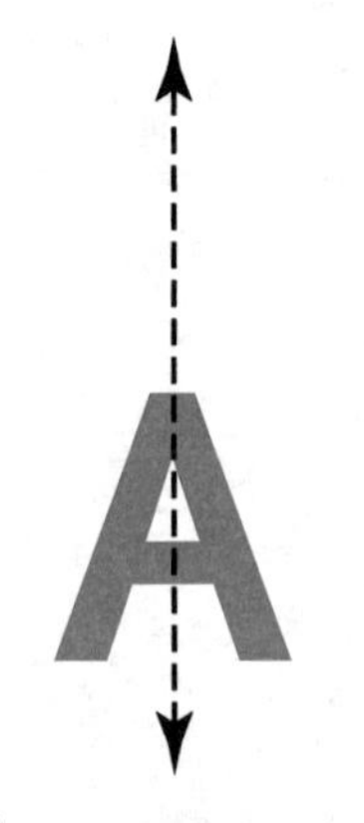

Line symmetry

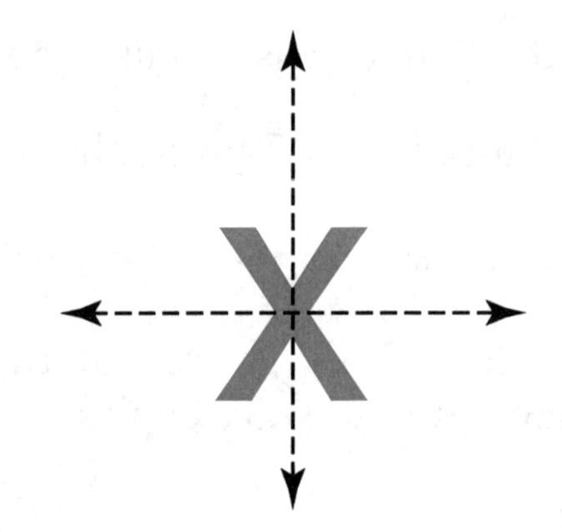

Line symmetry and rotation symmetry

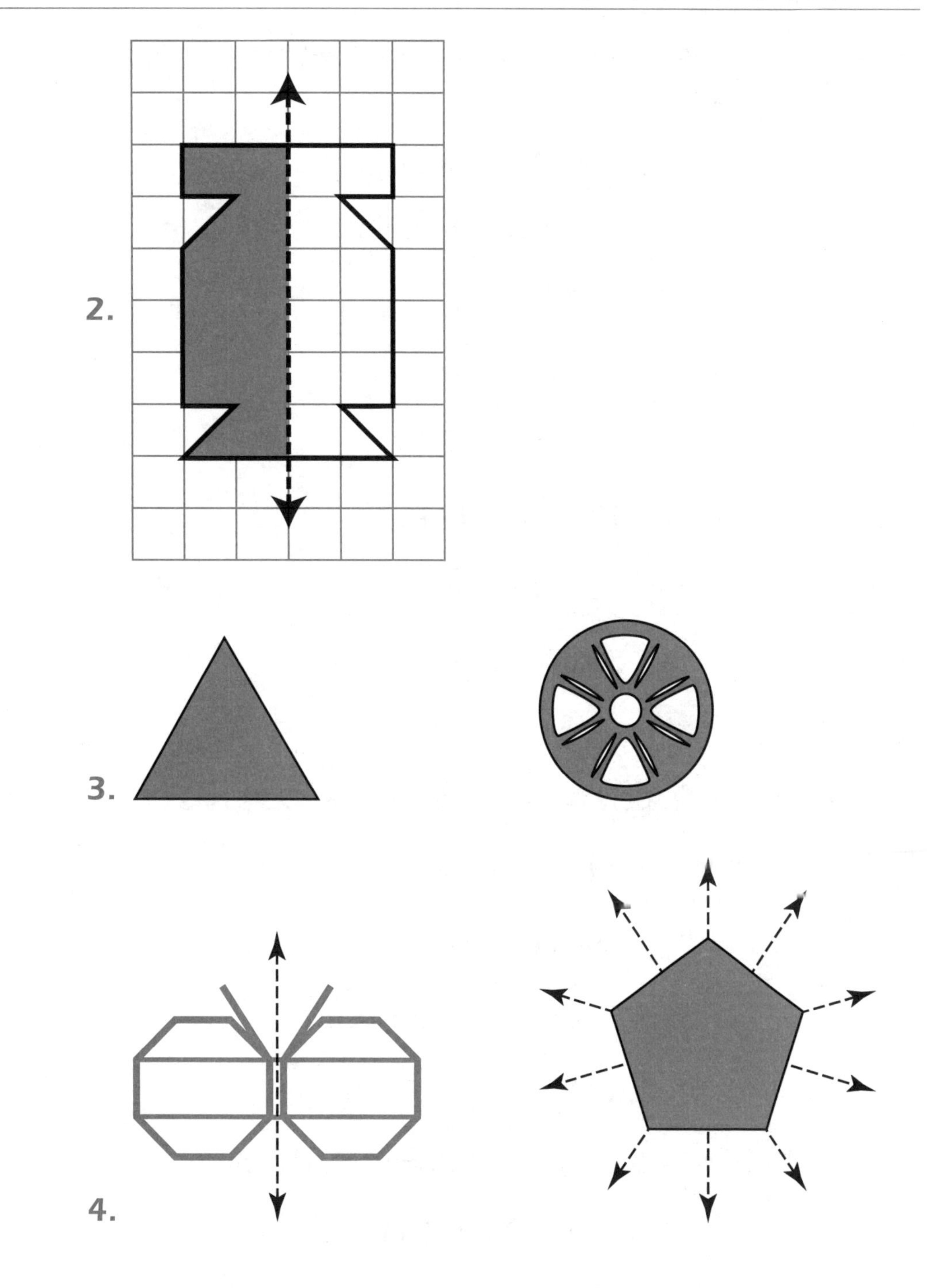

ANSWERS FOR TRANSFORMATIONS EXERCISES

1. Rotation
2. Reflection
3. Translation

ANSWERS FOR CONVERTING TRADITIONAL UNITS OF LENGTH EXERCISES

1. $9 \times 12 = 108$ in.
2. $5{,}280/1{,}760 = 3$ mi
3. $84/12 = 7$ ft
4. $9 \times 3 = 27$ ft
5. $159/3 = 53$ ft
6. $12 \times 12 = 144$ in
7. $78 \times 3 = 234$ ft
8. $5{,}280/5{,}280 = 1$ mi

ANSWERS FOR CONVERTING METRIC LENGTH EXERCISES

1. $4 \times 100 = 400$ cm
2. $9/100 = 0.09$ m
3. $1.5 \times 100 = 150$ cm
4. $545/100 = 5.45$ m
5. $1{,}000 \times 100 = 100{,}000$ cm
6. $10/100 = 0.1$ m
7. kilometers
8. centimeters
9. meters
10. millimeters

ANSWERS FOR MEASURING LENGTH EXERCISES

1. 7 cm
2. 2 in.
3. 10 cm

ANSWERS FOR CONVERTING THE U.S. TRADITIONAL WEIGHT EXERCISES

1. 96/16 = 6 lb
2. 18,000/2,000 = 9 T
3. 3.5 × 16 = 56 oz
4. 16 × 16 = 256 oz
5. 64/16 = 4 lb
6. 20,000/2,000 = 10 T
7. tons
8. pounds
9. ounces
10. pounds

ANSWERS FOR CONVERTING THE UNITS OF WEIGHT EXERCISES

1. 3,200/1,000 = 32 kg
2. 87 × 1000 = 87,000 g
3. 6,1000/1000 = 6.1 g
4. 240/1000 = 0.24 kg

5. $64.2 \times 1000 = 64{,}200$ mg

6. 19,570/1,000 = 19.57 kg

7. kilogram

8. grams

9. grams

10. milligram

ANSWERS FOR CONVERTING VOLUME EXERCISES

1. $9 \div 2 = 4\frac{1}{2}$ pt

2. $4 \times 4 = 16$ qt

3. $40 \div 8 = 5$ c

4. $4 \times 8 = 32$ fl oz

5. $18 \div 2 = 9$ qt

6. $6 \times 2 = 12$ pt

7. $5 \times 2 = 10$ c

8. $36 \div 4 = 9$ gal

ANSWERS FOR CONVERTING THE METRIC VOLUME EXERCISES

1. $9 \times 1{,}000 = 9{,}000$ mL

2. $4{,}000 \div 1{,}000 = 4$ L

3. $3{,}500 \div 1{,}000 = 3.5$ L

4. $6 \times 1{,}000 = 6{,}000$ mL

ANSWERS FOR PERIMETER EXERCISES

1. $2(10 + 6) = 32$ mi.

2. $2 \times 7 = 14$ cm

3. 2 ft = 24 in.; therefore $2(24 + 6) = 60$ in.

4. $(6 + 5 + 2 + 10 + 5 + 8) = 36$ ft

5. $2(3 + 4) = 14$ ft

6. $2(32 + 18) = 100$ in.

7. $5 + 4 + 10 + 2 = 21$ ft

ANSWERS FOR AREA EXERCISES

1. $7 \times 3 = 21$ m^2

2. $P = 8 \times 4 = 32$ cm; $A = 8 \times 8 = 64$ cm^2

3. $P = 2(1 + 2) = 6$ in.; $A = 2 \times 1 = 2$ in.^{2}

4. Since 15 units × 15 units = 225 square units, the side is 15 units in length.

5. C, a square of length 10 cm has perimeter $4 \times 10 =$ 40 cm.

6. $24 \times 12 = 288$ square yards

7. $80 \times 55 = 4{,}400$ square yards

8. Area $= (30 \times 5) + (25 \times 5) = 150 + 125 = 275$ square meters

9. 30 square feet
10. 20 square tiles

ANSWERS FOR COORDINATE GRID EXERCISES

1. Smiley face, (2, 3)
2. Star, (6, 1)
3. Basketball, (8, 9)
4. Hamburger, (7, 5)
5. Soccer ball, (4, 7)
6. Smiley face
7. Cookie
8. Apple
9. Soccer
10. Star
11. Triangle
12. Hamburger
13. Stop sign

14 to 21. See graph below.

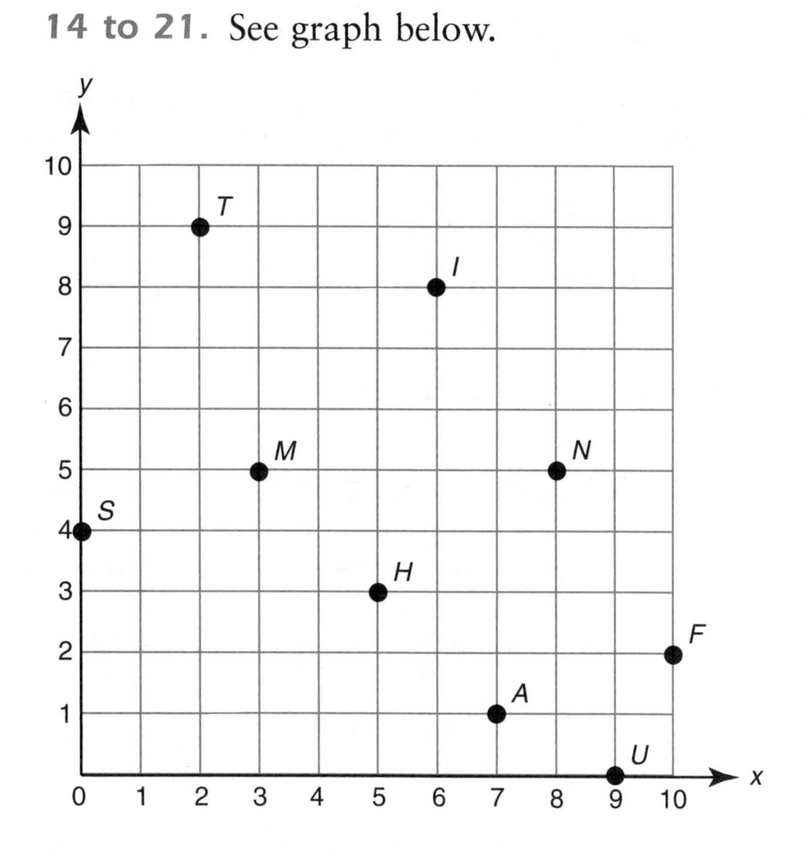

EXTENDED CONTRUCTED-RESPONSE ITEMS FOR CHAPTER 4

Question 1 You will receive full credit if:

- You correctly find that the cost of tiling the game room and closet is $1,200.
- You correctly find that the savings for tiling the game room and not the closet would be $200.
- You show work and explain how you got your answer.

Question 2 You will receive full credit if:

- You correctly find the fourth vertex of the parallelogram.
- You correctly list all coordinates of the parallelogram.
- You show your work for finding the perimeter of the parallelogram is 30 units.

Question 3 You will receive full credit if:

- You correctly draw in the two lines of symmetry.
- You correctly find the dimensions of the enlarged figure as 3 ft by $2\frac{1}{4}$ ft.
- You show your work for converting dimension from inches to feet.

CHAPTER 5

ANSWERS FOR PATTERNS EXERCISES

1. ABBABBABBABB (Answers will vary.)
2. ABCABCABCABC (Answers will vary.)
3. 1 dark, 1 light, 1 dark, 2 light, 1 dark, 3 light, 1 dark, 4 light, 1 dark, 5 light

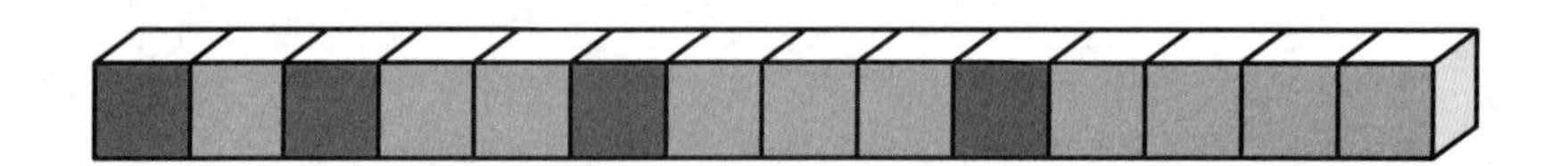

4. 1, 2, 1, 1, 2, 1, 1, 1, 2, 1, 1, 1, 1, 2, 1, 1, 1, 1, 1, 2
5. 2, 5, 8, 11, 14, 17, 20 (add 3)
6. 1, 2, 4, 8, 16, 32 (multiply by 2)
7. Ann, Brad, Carol, Daniel, Eleanor, Franklin (Names are in alphabetical order, names switch from girl, boy, girl, boy; and names increase in the number of letters used from 3 to 4 to 5 . . .)
8. 16, 37, 58, 79, 100 (add 21)
9. 1, 1, 2, 3, 5, 8, 13, 21, 34, 55 (add the sum of the previous two numbers to get the next number—Fibonacci sequence)

10. (triangle, pentagon, triangle, pentagon)

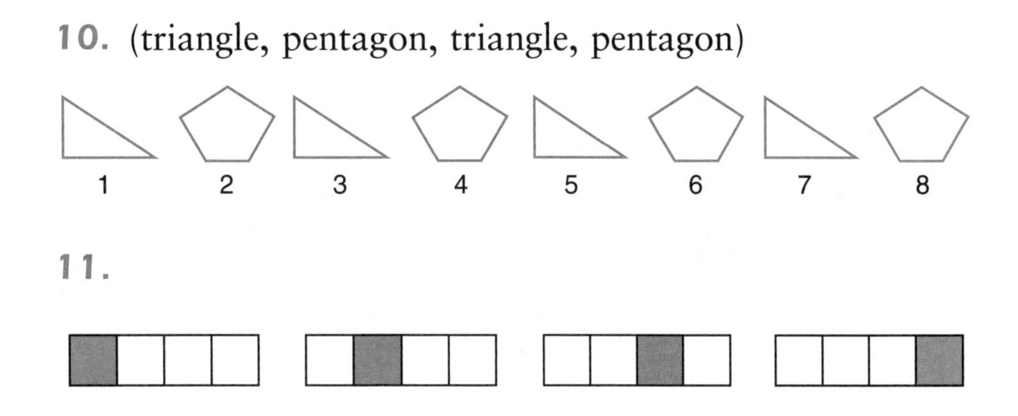

11.

12. 1, 4, 2, 8, . . .

The 4th, 8th, 12th, 16th, 20th, 24th, 28th, 32nd, 36th, 40th, 44th, 48th numbers are 8; therefore, the 49th number is 1 and the 50th number is 4.

13. 1, 1, 2, 4, 6, 18, 20, (×1, +1, ×2, +2, ×3, +3)

14. 70°F (5 p.m.), 67°F (6 p.m.), 64°F (7 p.m.), 61°F (8 p.m.), 58°F (9 p.m.), 55°F (10 p.m.), 52°F (11 p.m.)

15. O, T, T, F, F, S (One, Two, Three, Four, Five, Six)

16. 1, 2, 4, 8, 16 (×2); 1, 2, 4, 7, 11 (+1, +2, +3, +4), 1, 2, 4, 7, 28 (+1, ×2, +3, ×4)

ANSWERS FOR FUNCTIONS EXERCISES

1. Out = 2 × In

In	Out
2	4
3	6
11	22
27	**54**
9	18

2. Out = 3 × In + 1

In	Out
2	7
4	13
7	22
10	31
12	**37**
25	76

3. Out = 4 × In – 1

In	Out
One-eyed monster	3
Three-eyed monster	11
Four-eyed monster	15
Two-eyed monster	7

4. Out = (the number of letters in the In) – 1

In	Out
House	4
Cup	2
Writer	5
Mathematics	**10**
Elephant	**7**
Something	8
I	0

5.

Figure #	Toothpicks
1	4
2	7
3	10
4	13
5	16
6	19
n	$3n + 1$
33	100

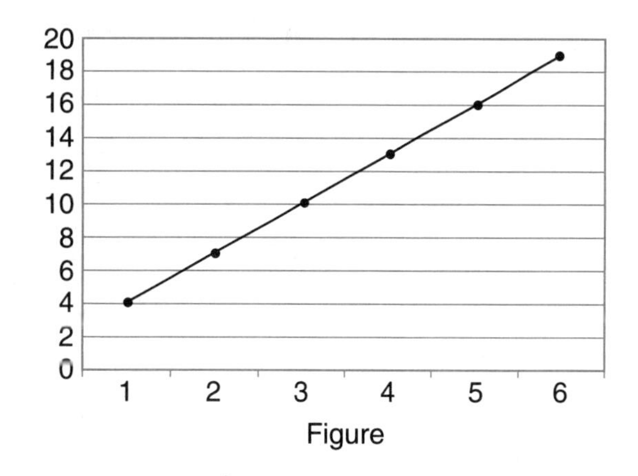

6.

Figure #	Toothpicks
1	3
2	5
3	7
4	9
5	11
6	13
n	$2n + 1$
50	101

7. In + Out = 60

In	10	17	28	40	59	x
Out	50	43	32	20	1	$60 - x$

8. Tyrone can buy 13 hamburgers for $19.50 and will receive 50 cents change. Graph the data.

Number of hamburgers	1	2	3	4	13	14
Cost	$1.50	$3.00	$4.50	$6.00	$19.50	$21.00

9. 10 + 7 − 3 = 14

10. C. A fish can swim 2 inches in 4 seconds.

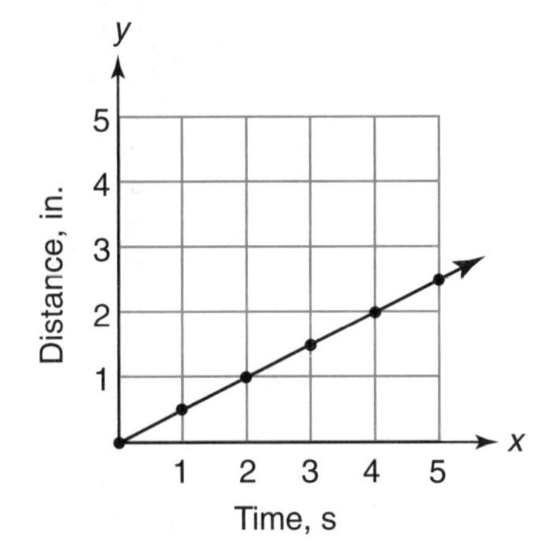

ANSWERS FOR NUMBER SENTENCES AND VARIABLES EXERCISES

1. Drummers = $F + 2$
2. Trumpet players = $F - 2$
3. Number in each row = $D \div 3$
4. Number of bags = $100 \div 5$
5. Number of boxes = $75 \div 15$

6. $65 + 30 + k = 155$

7. $5 + 5 + 5 - 3 = 12$

8. 9

9. 0

10. A. NO; B. YES; C. YES; D. YES; E. NO

11. A. Susan = $E + 5$; B. Maria = $2 \times E$; C. Donald = $E - 2$

12. A. $6 \times s = 30$; B. Kelly shoveled 5 driveways.

13. $$\begin{aligned} x + 3 &= 10 \\ -3 \quad &\; -3 \\ x &= 7 \end{aligned}$$

ANSWERS FOR GRAPHING LINEAR EQUATIONS FROM TABLES EXERCISES

1. $Y = X + 3$

X	Y
1	4
2	5
3	6

2.

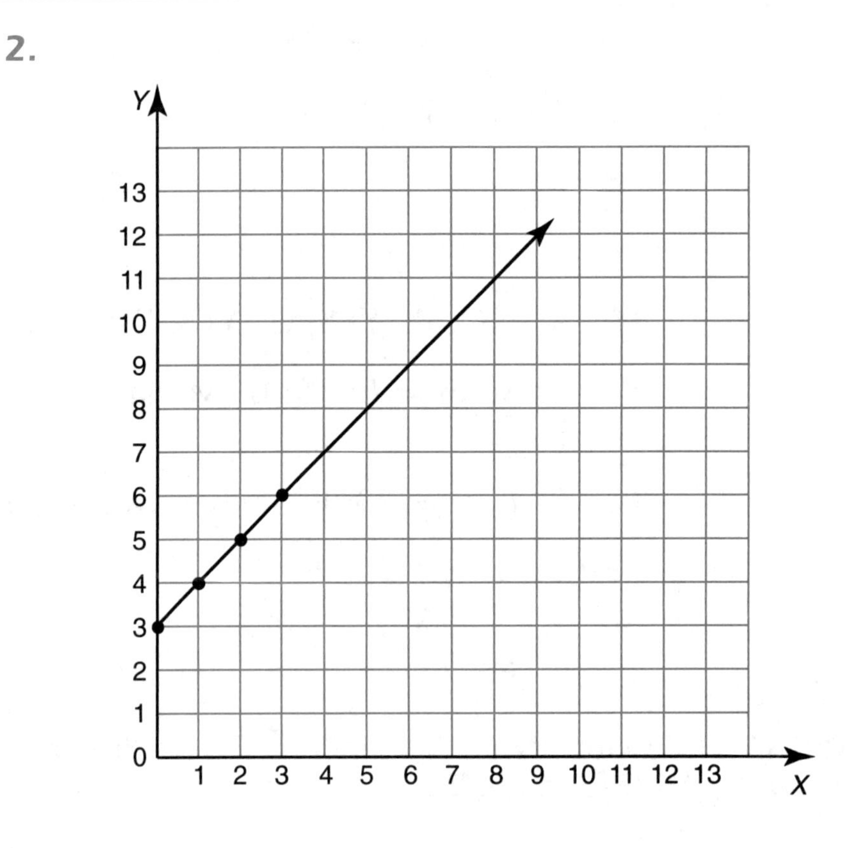

3. B. $Y = X + 3$

4. C.

X	0	2	4
Y	4	6	8

5. C. (2, 7)

6. $Y = 4X$.

X	Y
1	4
2	8
3	12

7.

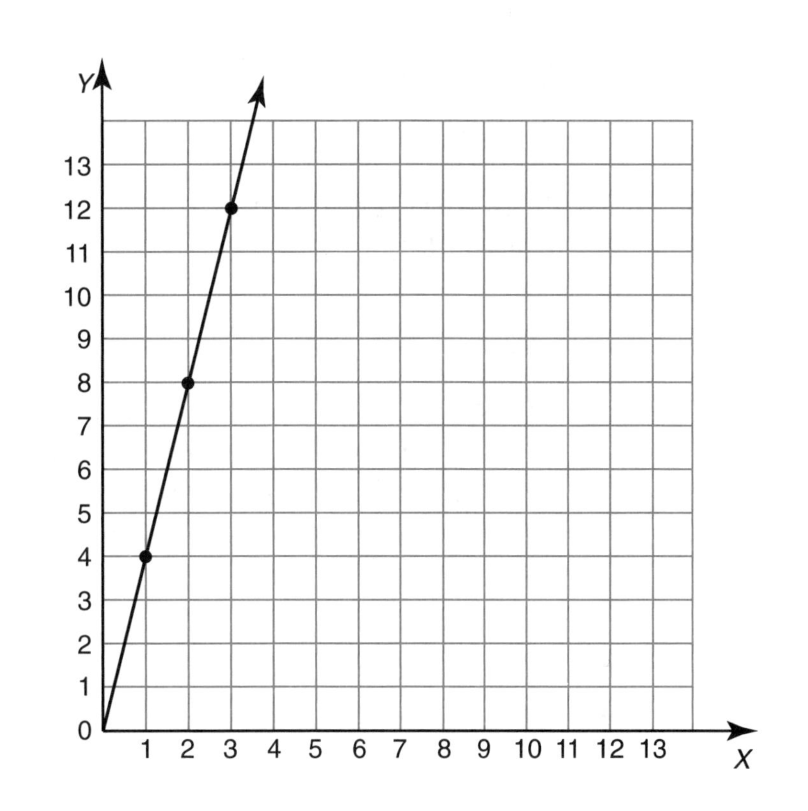

RUBRIC FOR EXTENDED CONSTRUCTED-RESPONSE ITEMS FOR CHAPTER 5

Question 1 You will receive full credit if:

- You correctly find the amount of money at the end of year 12 ($2,000).
- You explain the pattern to get this answer.
- You show your work for finding that Isabella can't afford the car (she will only have $3,250 at age 17).

Question 2 You will receive full credit if:

- You correctly find the next three figures in the pattern (ø Δ ø).
- You explain the pattern to get this answer.
- You correctly find Sarah's missing numbers (52, 54) and communicate that the rule is ×2, +2, ...

Question 3 You will receive full credit if:

- You correctly find that it takes 8 gold coins to balance a piggy bank. You show all work and explain how you got your answer.
- You correctly write a number sentence that represents the model.

Question 4

- You correctly complete the table with three ordered pairs that satisfy the equation $Y = 3X$. You show all work and explain how you got your answers.
- You correctly use the coordinates in your table to graph the line that represents the equation $Y = 3X$. You show all work and explain how you got your answers.

CHAPTER 6

ANSWERS FOR COLLECTING DATA EXERCISES

1. B. $\frac{7}{21} = \frac{1}{3}$

2. A. $\frac{14}{21} = \frac{2}{3}$

3. D. $\frac{21}{42} = \frac{1}{2}$

4. D. $\frac{12}{21} = \frac{4}{7}$

5. A. $\frac{6}{21} = \frac{2}{7}$

ANSWERS FOR COMPREHENDING DATA EXERCISES

1. Place the numbers in order: 2, 2, 3, 4, 5, 6, 6, 8, 9
   Range: $9 - 2 = 7$; Median: 5; Mean: $(8 + 4 + 9 + 2 + 5 + 6 + 3 + 6 + 2)/9 = 5$; Mode: 2, 6

2. Range: 8; Median: 7; Mean: 6; Mode: 9
3. Range: 11; Median: 5; Mean: 5.5; Mode: 1
4. C. 14
5. D. 22
6. B. 19
7. D. 2, 3
8. C. 3
9. C. 3
10. B. 5

ANSWERS FOR COMPARING DATA—BAR GRAPH EXERCISES

1. B. 3
2. A. 20
3. B. Computer
4. C. 5
5. D. 5
6. C. 2
7. 12 – 2 = 10
8. Telephone
9. 30 students
10. C. 100
11. B. May
12. D. 10

13. C. increased

14. A. June

ANSWERS FOR COMPARING DATA—LINE GRAPHS EXERCISES

1. B. 100
2. D. $500 (appears twice)
3. C. $200 ($500 – $300)
4. B. Saturday $900—highest point on the graph
5. D. Monday
6. The trend is that Mandy's Handy Candy Shop seems to earn more money on the weekends. Answers may vary.

ANSWERS FOR COMPARING DATA—CIRCLE GRAPHS EXERCISES

1. C. Spend time with family—50%
2. A. Study/school work—5%
3. B. Hang out with friends—20%
4. C. 15% (20 – 5)
5. B. 10 ($20 \times 50\% = 10$)

ANSWERS FOR INTUITIVE PROBABILITY EXERCISES

1. C. Less likely, there is only one green section.
2. A. Likely, there are many sections with the number 1.
3. The red lollipops have a greater probability of being chosen: P(red) $= \frac{3}{9} = \frac{1}{3}$.

4. The green, yellow, and blue lollipops have the same probability of being chosen.

$$P(\text{green}) = P(\text{yellow}) = P(\text{blue}) = \frac{2}{9}.$$

ANSWERS FOR EXPERIMENTAL PROBABILITY EXERCISES

1.

Favorable Outcomes of a Six-Sided Die	Frequency of Times Number Is Rolled	Experimental Probability Rolling a Number
1	2	$P(1) = \frac{2}{15}$
2	1	$P(2) = \frac{1}{15}$
3	0	$P(3) = \frac{0}{15}$
4	4	$P(4) = \frac{4}{15}$
5	3	$P(5) = \frac{3}{15} = \frac{1}{5}$
6	5	$P(6) = \frac{5}{15} = \frac{1}{3}$

2.

Favorable Outcomes	Frequency of Outcome	Experimental Probability
Head	8	$P(H) = \frac{8}{12} = \frac{2}{3}$
Tail	4	$P(T) = \frac{4}{12} = \frac{1}{3}$

ANSWERS FOR THEORETICAL PROBABILITY EXERCISES

1. There are 7 boy names: Gary, James, Matt, Eric, Mike, Will, Kevin. There are 12 possible outcomes.

$$P \text{ (boy's name)} = \frac{7}{12}$$

2. There are 5 girl names: Evelyn, Jenny, Jill, Maria, Kesha.

$$P \text{ (girl's name)} = \frac{5}{12}$$

3. There are 2 names that begin with a vowel: Eric and Evelyn.

$$P \text{ (name beginning with vowel)} = \frac{2}{12} = \frac{1}{6}$$

4. **B.** There are 3 even numbers: 2, 4, 6. There are 6 possible outcomes.

$$P \text{ (even number)} = \frac{1}{2}$$

5. **D.** There are 2 multiples of the number 3: 3, 6.

$$P \text{ (multiples of 3)} = \frac{2}{6} = \frac{1}{3}$$

6. **C.** The only number greater than 5 is 6.

$$P \text{ (number greater than 5)} = \frac{1}{6}$$

7. **A.** There are 6 checkered marbles. There are a total of 15 possible outcomes.

$$P \text{ (checkered)} = \frac{6}{15} = \frac{2}{5}$$

8. **D.** There are 3 white marbles.

$$P \text{ (white)} = \frac{3}{15} = \frac{1}{5}$$

9. C. There is 1 striped marble.

$$P\text{ (striped)} = \frac{1}{15}$$

10. B. There are 5 dark marbles.

$$P\text{ (dark)} = \frac{5}{15} = \frac{1}{3}$$

ANSWERS FOR MULTIPLICATION PRINCIPLE OF COUNTING EXERCISES

1. D. 80 days

2. A. 12

3. $4 \times 3 = 12$

4. Jim, Veronica, Mia

   Jim, Mia, Veronica

   Veronica, Jim, Mia

   Veronica, Mia, Jim

   Mia, Jim, Veronica

   Mia, Veronica, Jim

   The probability that Jim or Veronica will come in first is $\frac{4}{6} = \frac{2}{3}$.

5. Table of Possible Sums of a Roll of 2 Dice

	1	2	3	4	5	6
1	2	3	4	5	6	7
2	3	4	5	6	7	8
3	4	5	6	7	8	9
4	5	6	7	8	9	10
5	6	7	8	9	10	11
6	7	8	9	10	11	12

There are 36 possible outcomes. Out of the 36 outcomes only six of the outcomes have a sum of 7.

1-6, 2-5, 3-4, 4-3, 5-2, 6-1

There are 6 ways to roll two dice and get a sum of seven.

ANSWERS FOR VERTEX-EDGE GRAPHS EXERCISES

1. Pet Store–Playground–Home, 10 + 4 = 14 units
2. **C.** Home–Playground–School–Pet Store, 4 + 8 + 7 = 19 units
3. School–Playground–Home–Pizza Parlor, 8 + 4 + 3 = 15 units
4. **B.** 7 vertices
5. **A.** 9 edges
6. **C.** 9

EXTENDED–CONSTRUCTED RESPONSE ITEMS FOR CHAPTER 6

Question 1 You will receive full credit if:

- You correctly find that 20 fifth graders said that Game System A was their most recent video game system and 5 fifth graders said that Game System D was their most recent video game system.
- You correctly find that 15 more fifth graders said that Game System A was their most recent video game console.
- You show work and explain how you found that 15 more fifth graders said that Game System A was their most recent video game console.

Question 2 You will receive full credit if:

- You correctly find the 15 outcomes are prime numbers.
- You correctly find that the probability of winning is $\frac{15}{36} = \frac{5}{12}$.
- You show your work and explain that the game is not fair because players do not have an equal chance of winning. The probability would have to be $\frac{18}{36}$ or $\frac{1}{2}$.

First Number Cube

Second Number Cube	1	2	3	4	5	6
1	2	3	4	5	6	7
2	3	4	5	6	7	8
3	4	5	6	7	8	9
4	5	6	7	8	9	10
5	6	7	8	9	10	11
6	7	8	9	10	11	12

Question 3 You will receive full credit if:

- You use appropriate title, label, and scale.
- You correctly display data on graph: Water Rides—3 students, Rollercoaster—7 students, Anti-gravity Rides—3 students, Bumper Cars—4 students, and Other—1 student.
- Answers may vary but you correctly make a comparison of data.

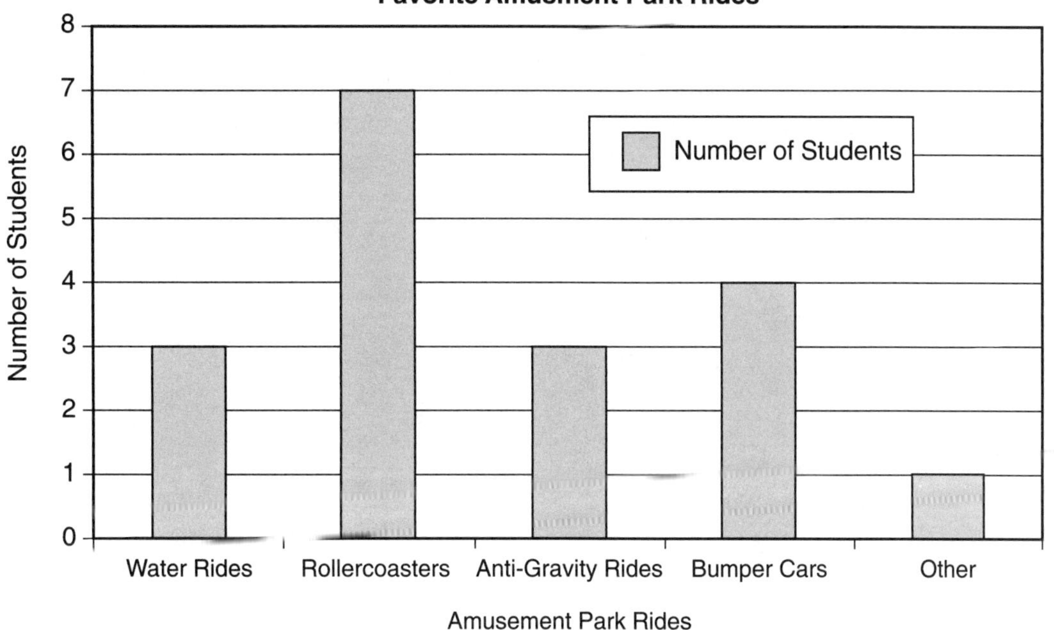

ANSWER SHEET: SAMPLE TEST 1

Section 1
Short Constructed-Response Questions

1. ______________________________

2. ______________________________

3. ______________________________

4. ______________________________

5. ______________________________

6. ______________________________

Section 2
Multiple-Choice Questions

7. Ⓐ Ⓑ Ⓒ Ⓓ
8. Ⓐ Ⓑ Ⓒ Ⓓ
9. Ⓐ Ⓑ Ⓒ Ⓓ
10. Ⓐ Ⓑ Ⓒ Ⓓ
11. Ⓐ Ⓑ Ⓒ Ⓓ
12. Ⓐ Ⓑ Ⓒ Ⓓ
13. Ⓐ Ⓑ Ⓒ Ⓓ
14. Ⓐ Ⓑ Ⓒ Ⓓ
15. Ⓐ Ⓑ Ⓒ Ⓓ
16. Ⓐ Ⓑ Ⓒ Ⓓ
17. Ⓐ Ⓑ Ⓒ Ⓓ
18. Ⓐ Ⓑ Ⓒ Ⓓ

Extended Constructed-Response Questions

19.

Section 3
Multiple-Choice Questions

20. Ⓐ Ⓑ Ⓒ Ⓓ
21. Ⓐ Ⓑ Ⓒ Ⓓ
22. Ⓐ Ⓑ Ⓒ Ⓓ
23. Ⓐ Ⓑ Ⓒ Ⓓ
24. Ⓐ Ⓑ Ⓒ Ⓓ
25. Ⓐ Ⓑ Ⓒ Ⓓ
26. Ⓐ Ⓑ Ⓒ Ⓓ
27. Ⓐ Ⓑ Ⓒ Ⓓ
28. Ⓐ Ⓑ Ⓒ Ⓓ
29. Ⓐ Ⓑ Ⓒ Ⓓ
30. Ⓐ Ⓑ Ⓒ Ⓓ
31. Ⓐ Ⓑ Ⓒ Ⓓ

Extended Constructed-Response Questions

32.

Section 4

Multiple-Choice Questions

33. Ⓐ Ⓑ Ⓒ Ⓓ
34. Ⓐ Ⓑ Ⓒ Ⓓ
35. Ⓐ Ⓑ Ⓒ Ⓓ
36. Ⓐ Ⓑ Ⓒ Ⓓ
37. Ⓐ Ⓑ Ⓒ Ⓓ
38. Ⓐ Ⓑ Ⓒ Ⓓ
39. Ⓐ Ⓑ Ⓒ Ⓓ
40. Ⓐ Ⓑ Ⓒ Ⓓ
41. Ⓐ Ⓑ Ⓒ Ⓓ
42. Ⓐ Ⓑ Ⓒ Ⓓ
43. Ⓐ Ⓑ Ⓒ Ⓓ

Extended Constructed-Response Question

44.

SAMPLE TEST #1

SECTION 1

SHORT CONSTRUCTION RESPONSE

Directions: For each question, write the correct answer on the line provided on the answer sheet. You may NOT use a calculator for this section.

1. The table below shows the number of family members in each person's family. What is the number of members in Monique's family if the mean of all the families was 5?

Student	Family Members
Juan	4
Karlton	6
Mario	4
Frankie	7
Susanna	5
Monique	?

2. Louie made 17 bag lunches for the school outing. If Louie had made 4 more lunches, he would have made exactly 3 times as many bag lunches as Marc did. How many lunches did Marc make?

3. In Sarah's fifth grade class, 4 students out of 25 brought their lunch to school today. What percent of students brought their lunch to school?

4. Seven apples are sold in a bag. Joshua buys six bags. How many apples does he have?

5. Estimate the measure of the angle below.

6. Julia has many toys in her toy chest. 25% of the toys are dolls. What fraction of the toys are NOT dolls.

SECTION 2

MULTIPLE CHOICE

Directions: Darken the letter of the best answer on the answer sheet. You may use a calculator for this section.

7. The design below was drawn by Richardo.

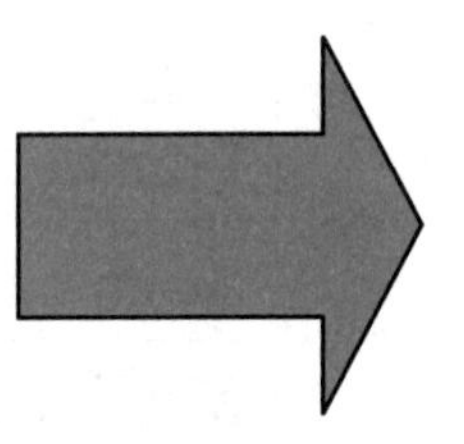

Which of the following represents the design flipped over a horizontal line?

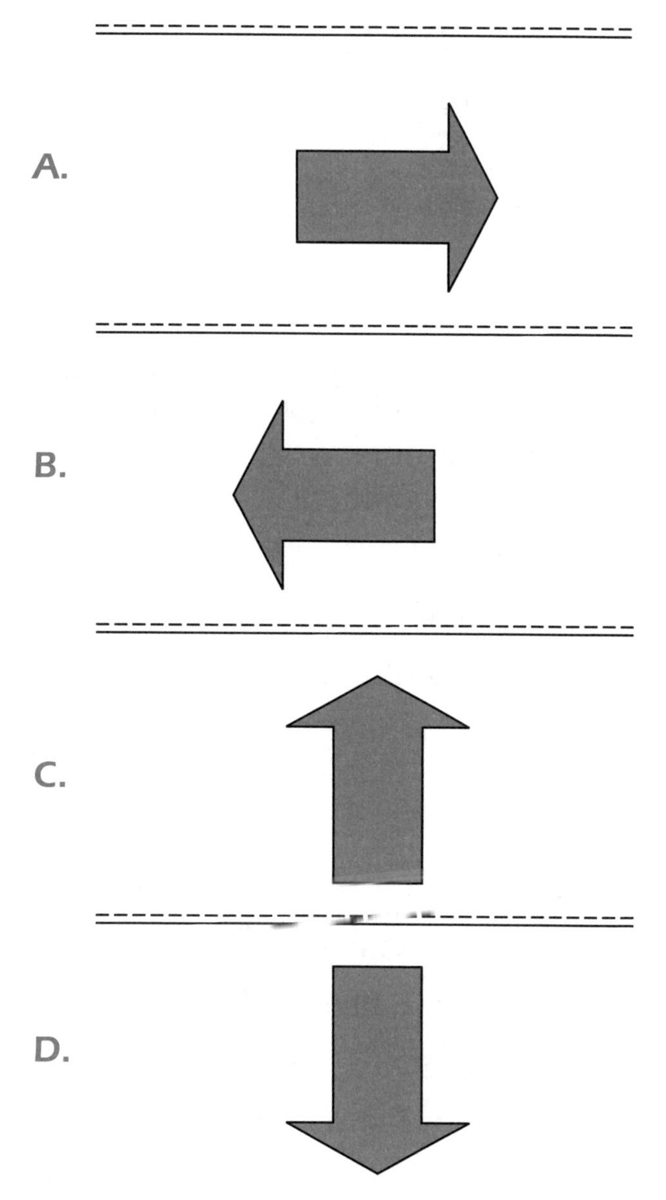

8. What are the next two numbers in the sequence: 2, 3, 5, 8, 12, 17, _____ , _____?

A. 22, 27

B. 23, 29

C. 23, 30

D. 27, 37

9. A bottle of mustard weighs 1.5 pounds; how many ounces does the bottle weigh?

   A. 12
   B. 16
   C. 20
   D. 24

10. Marge arranged the four numbers 5, 8, 9, and 1 to make the largest number that she could. What is the largest number she could make using each of the above numbers just once?

   A. 9,815
   B. 9,851
   C. 1,859
   D. 981

11. The best estimate of 67,893–21,890 is _____.

   A. less than 4,000
   B. more than 4,000
   C. more than 40,000
   D. less than 40,000

12. Given the number 7.56, if the 5 is changed to the number 2, then _____.

   A. the new number is bigger by 3 ones
   B. the new number is smaller by 3 tenths
   C. the new number is smaller by 3 hundredths
   D. the new number is smaller by 3 ones

13. Jose received the following five grades on his math tests: 95, 90, 85, 80, and 75. What was the average of his grades?

A. 80

B. 82.5

C. 85

D. 90

14. A dog's pen is drawn below with an area of 500 square meters. The length of the pen is 20 meters; what is its width?

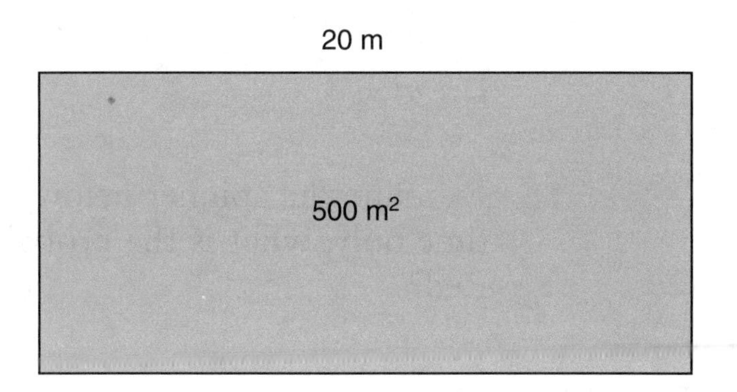

A. 20 m

B. 25 m

C. 230 m

D. 480 m

15. Macala bought bathroom tiles which were 6 in. by 6 in. How many will be needed to cover an area of 720 square inches?

A. 5

B. 10

C. 20

D. 120

16. Given the number sentence $10 - b = 3$. What is the value of b?

    A. 13

    B. 10

    C. 7

    D. 3

17. If m represents a number, which of the following means 3 less than the number?

    A. $3 - m$

    B. $m - 3$

    C. $3 \times m$

    D. $m \div 3$

18. Examine the spinner below. If the spinner is spun one time only, what is the probability that it will land on Red?

    A. $\frac{1}{6}$

    B. $\frac{1}{4}$

    C. $\frac{1}{3}$

    D. $\frac{1}{2}$

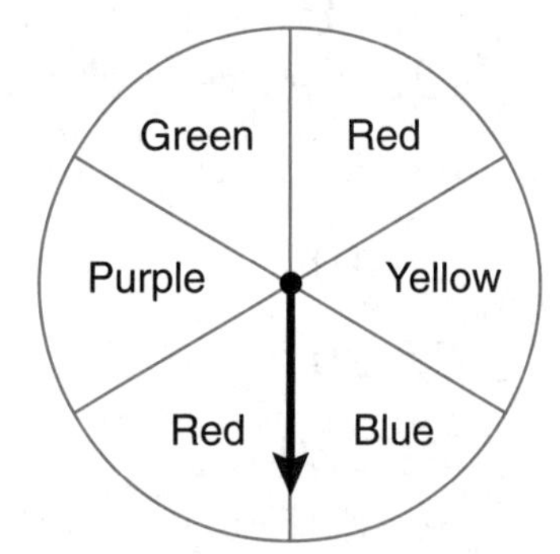

EXTENDED CONSTRUCTED-RESPONSE QUESTIONS

Directions: Write your response in the space provided on the answer sheet. Answer the question as completely as possible. You may use a calculator for this section.

19. The Anybody Can Play Football League came up with a new scoring system: 3 points for a field goal and 8 points for a touchdown. In this new league, players can score with a touchdown, field goal, or any combination of the two. One of the players figured out that not all scores are possible.

- Highlight all the multiples of 3 and 8 on the chart.
- Find all scores that are not possible.
- Show work and explain answers.

1	2	3	4	5	6	7	8	9	10
11	12	13	14	15	16	17	18	19	20
21	22	23	24	25	26	27	28	20	30
31	32	33	34	35	36	37	38	39	40
41	42	43	44	45	46	47	48	49	50
51	52	53	54	55	56	57	58	59	60
61	62	63	64	65	66	67	68	69	70
71	72	73	74	75	76	77	78	79	80
81	82	83	84	85	86	87	88	89	90
91	92	93	94	95	96	97	98	99	100

On the number line below, plot and label points for the following numbers: $\frac{3}{5}$ and 0.8.

- List three different fractions or decimals that are greater than $\frac{3}{5}$ and less than 0.8.
- Explain how you know that the three fractions/decimals are greater than $\frac{3}{5}$ and less than 0.8.

SECTION 3

MULTIPLE CHOICE

Directions: Darken the letter of the best answer on the answer sheet. You may use a calculator for this section.

20. Richardo has a rock collection. He gave 12 rocks to his brother and 3 rocks to his sister. Richardo now has 20 rocks left. How many did he start with?

A. 5

B. 12

C. 23

D. 35

21. The table below shows attendance at seven NJ Devils hockey games for the month of March. What is the best estimate of the total attendance for March?

Date	Attendance
March 3	18,235
March 4	19,152
March 12	18,789
March 15	17,920
March 18	18,193
March 21	19,237
March 25	19,890

A. About 20,000

B. About 140,000

C. About 200,000

D. About 1,000,000

22. Given the following fractions all with a numerator of one: $\frac{1}{2}, \frac{1}{3}, \frac{1}{4}, \frac{1}{5}$. Which statement is correct about these fractions?

A. Since the numerator is always 1, the fractions are equal.

B. As the denominator increases, the fraction increases.

C. As the denominator increases, the fraction decreases.

D. The fractions are getting closer to 1.

23. Which of the following is not a parallelogram?

A. Trapezoid

B. Square

C. Rectangle

D. Rhombus

24. How many lines of symmetry does the figure below have?

X

A. 0 B. 1 C. 2 D. 3

25. Which of the following is the best estimate of the solution to 29×101?

A. 300

B. 3,000

C. 30,000

D. 300,000

26. Jacob's pet snake measures 72.18 centimeters. What is the snake's length rounded to the nearest tenth of a centimeter?

A. 72

B. 72.1

C. 72.18

D. 72.2

27. A recipe calls for $\frac{1}{2}$ cup of sugar to make brownies. Which of the following amounts is equivalent to $\frac{1}{2}$ cup?

A. $\frac{2}{3}$ cup

B. $\frac{3}{4}$ cup

C. $\frac{3}{6}$ cup

D. $\frac{4}{9}$ cup

28. In Mrs. Jackson's class, 5 students have red hair, 3 students have brown hair, and 2 students have blond hair. What is the probability that Mrs. Jackson selects a student with blond hair to read a passage from the textbook?

A. $\frac{1}{10}$

B. $\frac{2}{10}$

C. $\frac{3}{10}$

D. $\frac{5}{10}$

29. The table below shows the number of miles that Johanna ran each day last week. What was the mean (average) number of miles she ran each day?

Date	Miles
Monday	5
Tuesday	7
Wednesday	8
Thursday	10
Friday	8
Saturday	7
Sunday	4

A. 10
B. 8
C. 7
D. 6

30. Mimi has a bag of marbles. Inside the bag are 10 blue marbles, 6 green marbles, and 4 white marbles. If she reaches inside the bag and selects one, what is the probability that it is NOT green?

A. $\frac{6}{20}$

B. $\frac{10}{20}$

C. $\frac{14}{20}$

D. $\frac{16}{20}$

31. Given the following numbers: 2, 3, 5, 7, and 9, what is the fraction of the numbers that are prime?

A. 0

B. $\frac{3}{5}$

C. $\frac{4}{5}$

D. 1

EXTENDED CONSTRUCTED-RESPONSE QUESTIONS

Directions: Write your response in the space provided on the answer sheet. Answer the question as completely as possible. You may use a calculator for this section.

32. The table below shows how many minutes Jilian reads for the 100 Book Challenge each day.

100 Book Challenge

Number of Days, n	Total Minutes Read, m
1	15
2	30
3	45
4	60
5	
6	
7	
n	

- Complete the missing numbers in the table and write a rule that you could use to help determine the number of minutes read in *n* days.
- How many minutes were read in 40 days? Show work and explain answer.
- How many hours did Jillian read in 40 days? Show work and explain answer.

SECTION 4

MULTIPLE CHOICE

Directions: Darken the letter of the best answer on the answer sheet. You may use a calculator for this section.

33. There are 175 balls in a large box. Some of the balls are white and some of the balls are red. If 73 of the balls are white, which shows how to find the number of red balls?

A. $175 + 73$

B. $175 - 73$

C. 175×73

D. $175 \div 73$

34. Austin has 52 toys and his sister, Jordyn, has 65 toys. Austin and Jordyn want to combine their toys and then share them with nine friends. How many toys will each of the nine friends get?

A. 117

B. 65

C. 13

D. 9

35. Four brothers were measuring their heights. Alex is shorter than Brendan. Alex is taller than Charles. David's height is between Brendan's height and Alex's height. Who is the shortest person?

A. Alex

B. Brendan

C. Charles

D. David

36. Examine the clock below. What kind of angle is formed by its hands?

A. Right

B. Acute

C. Obtuse

D. Left

37. Ryan has 128 hats, which is 18 more than Stuart has. Which equation should be used to find h, the number of hats that Stuart has?

A. $h = 128 + 18$

B. $h = 128 - 18$

C. $h = 128/18$

D. $h = 128 \times 18$

38. Which table represents values of x and y such that $y = x - 2$?

A.

x	y
4	6
6	8
8	10

B.

x	y
4	2
6	4
8	6

C.

x	y
4	8
6	12
8	16

D.

x	y
4	4
6	6
8	8

39. George scored the following numbers of pins in 5 games of bowling: 99, 86, 112, 105, 134. What is the median of these numbers?

A. 86

B. 99

C. 105

D. 112

40. Find the next number in the sequence: 7, 10, 14, 19, 25, . . .

A. 31

B. 32

C. 33

D. 34

41. The following students are on a basketball team: Peter, Harry, Paul, Tamika, Juan, and Perry. If one student is the captain of the team, what is the probability that his name starts with a P?

A. $\frac{1}{6}$

B. $\frac{1}{3}$

C. $\frac{1}{2}$

D. 1

42. The table below shows the race times of four students. Which student has the fastest race time?

Name	Race time
Karen	11.235
Charissa	11.29
Sarah	11.3
Michelle	11.199

A. Karen

B. Charissa

C. Sarah

D. Michelle

43. Austin has four crayons—a red one, a blue one, a green one, and a yellow one. He is using only two crayons to color his poster. How many different pairs of crayons could he use?

A. 2

B. 4

C. 6

D. 8

EXTENDED CONSTRUCTED-RESPONSE QUESTION

Directions: Write your response in the space provided on the answer sheet. Answer the question as completely as possible. You may use a calculator for this section.

44. In the Summer Theatre Production of *Holka Polka*, the cast consists of the following witches. The following table includes the number of lines that each witch has.

Witch	Number of Lines
Hilda	18
Gilda	24
Gandolt	14
Scandolt	14
Shurka	7
Snorz	17
Sweet	12
Zorka	14
Zoom	7
Splenda	13

- Find the mean of the lines.
- Find the median of the lines.

- If we added a new witch, Brenda, with 48 lines, how would the mean change? Show work and explain answer.

ANSWERS to SAMPLE TEST #1

1. 6 families × mean of 5 = 30 members. 30 – (4 + 6 + 4 + 7 + 5) = 4.
2. 17 + 4 = 21 = 3 × 7. Thus Marc made 7 lunches.
3. 4/25 = 16%
4. 7 × 6 = 42
5. 80 degrees (Any answer from 75 to 85 is acceptable.)
6. $\frac{3}{4} = \frac{75}{100} = .75$ (All are acceptable.)
7. A.
8. C. Add one, add two, add three, ...
9. D. 1.5 × 16 = 24
10. B. 9,851
11. C. 68,000 – 22,000 = 46,000
12. B. 7.26 is .3 less than 7.56
13. C. (95 + 90 + 85 + 80 + 75)/5 = 85
14. B. 500/20 = 25
15. C. 720/36 = 20
16. C. 10 – 7 = 3
17. B. m – 3

18. C. $\frac{2}{6} = \frac{1}{3}$

19. You will receive full credit if:

- You correctly identify the multiples of 3: 3, 6, 9, 12, 15, 18, 21, 24, 27, 30, 33, 36, 39, 42, 45, 48, 51, 54, 57, 60, 63, 66, 69, 72, 75, 78, 81, 84, 87, 90, 93, 96, and 99.
- You correctly identify the multiples of 8: 8, 16, 24, 32, 40, 48, 56, 64, 72, 80, 88, and 96.
- You correctly determine that 1, 2, 5, 7, 10, and 13 are not possible scores, because there are no combinations of 3 and 8 that give you those scores even if you add 3 or 8 to the multiples.

20. D. $35 - (12 + 3) = 20$

21. B. $20{,}000 \times 7 = 140{,}000$

22. C. As the denominator increases, the fraction decreases.

23. A. Trapezoids have only one pair of parallel sides.

24. C. 2—horizontal and vertical

25. B. $30 \times 100 = 3{,}000$

26. B. 72.1

27. C. $\frac{3}{6} = \frac{1}{2}$

28. B. $\frac{2}{10}$

29. C. $(5 + 7 + 8 + 10 + 8 + 7 + 4)/7 = 7$

30. **C.** $\frac{14}{20}$

31. **C.** $\frac{4}{5}$, 2, 3, 5, and 7 are prime.

32. You will received full credit if:

- You correctly fill in the table (5 days, 75 minutes; 6 days, 90 minutes; 7 days, 105 minutes).
- You correctly write the rule as $n \times 15$.
- You correctly support findings that Jillian reads 600 minutes in 40 days and 10 hours in 40 days.

33. **B.** 175 – 73

34. **C.** (52 + 65)/9 = 13

35. **C.** Charles (Brendan, David, Alex, Charles)

36. **A.** Right

37. **B.** $h = 128 - 18 = 110$ hats for Stuart

38. **D.**

39. **C.** 105. In order: 86, 99, 105, 112, 134

40. **B.** 32 (add one, add two, add three, . . .)

41. **C.** $\frac{3}{6} = \frac{1}{2}$

42. **D.** Michelle (smallest number is fastest)

43. **C.** 6 RB, RG, RY, BG, BY, GY

44. You will receive full credit if:

- You correctly find that the mean number of witch lines is 14 (140 total lines ÷10 witches).
- You correctly find that the mode is 14 lines.
- You correctly find and support findings that the mean would change to approximately 17 or 17.1.

ANSWER SHEET: SAMPLE TEST 2

Section 1
Short Constructed-Response Questions

1. ______________________________

2. ______________________________

3. ______________________________

4. ______________________________

5. ______________________________

6. ______________________________

Section 2
Multiple-Choice Questions

7. Ⓐ Ⓑ Ⓒ Ⓓ
8. Ⓐ Ⓑ Ⓒ Ⓓ
9. Ⓐ Ⓑ Ⓒ Ⓓ
10. Ⓐ Ⓑ Ⓒ Ⓓ
11. Ⓐ Ⓑ Ⓒ Ⓓ
12. Ⓐ Ⓑ Ⓒ Ⓓ
13. Ⓐ Ⓑ Ⓒ Ⓓ
14. Ⓐ Ⓑ Ⓒ Ⓓ
15. Ⓐ Ⓑ Ⓒ Ⓓ
16. Ⓐ Ⓑ Ⓒ Ⓓ
17. Ⓐ Ⓑ Ⓒ Ⓓ
18. Ⓐ Ⓑ Ⓒ Ⓓ

Extended Constructed-Response Questions

19.

Section 3
Multiple-Choice Questions

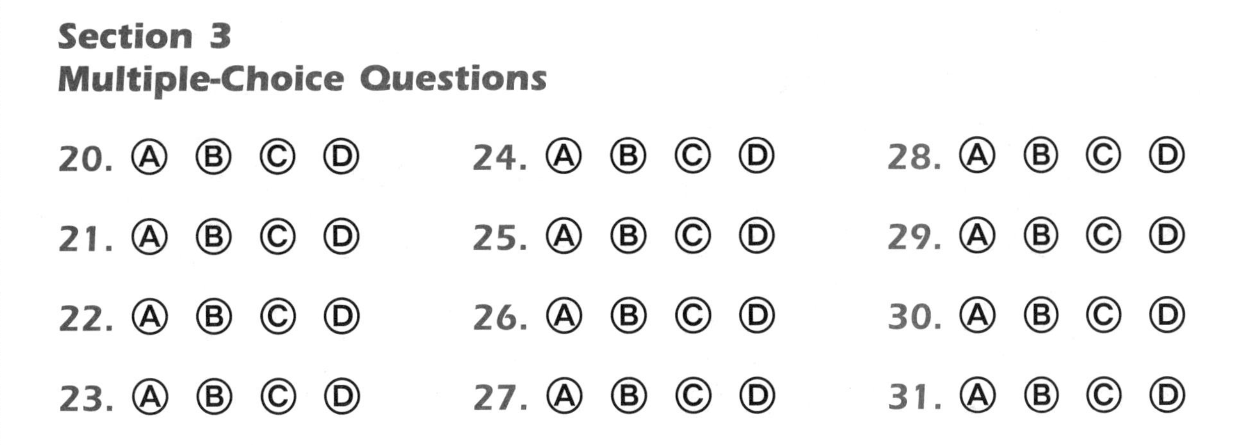

Extended Constructed-Response Questions

32.

Section 4
Multiple-Choice Questions

33. Ⓐ Ⓑ Ⓒ Ⓓ
34. Ⓐ Ⓑ Ⓒ Ⓓ
35. Ⓐ Ⓑ Ⓒ Ⓓ
36. Ⓐ Ⓑ Ⓒ Ⓓ
37. Ⓐ Ⓑ Ⓒ Ⓓ
38. Ⓐ Ⓑ Ⓒ Ⓓ
39. Ⓐ Ⓑ Ⓒ Ⓓ
40. Ⓐ Ⓑ Ⓒ Ⓓ
41. Ⓐ Ⓑ Ⓒ Ⓓ
42. Ⓐ Ⓑ Ⓒ Ⓓ
43. Ⓐ Ⓑ Ⓒ Ⓓ

Extended Constructed-Response Questions

44.

SAMPLE TEST #2

SECTION 1 SHORT CONSTRUCTION RESPONSE

Directions: For each question, write the correct answer on the line provided on the answer sheet. You may NOT use a calculator for this section.

1. Given the following clues, find the mystery number. It is an even number. It is a multiple of 3. It is divisible by 5. It is less than 40.

2. The fastest tennis serve ever measured was one of 163.6 mph by William Tatem in 1931. What was the speed rounded to the nearest whole number?

3. In June of 2008, Jamaica Bolt set the world record for the 100-meter race. His time was 9 seconds and 72 hundredths. Write this number as a decimal.

4. What number, b, would make the number sentence true? $28 + b = 50$.

5. Renaldo scores on his four math tests were 100, 90, 95, and 95. What was the average score of Renaldo's tests?

6. A garden needs fencing around its borders. The garden is in the shape of a rectangle with length = 20 ft and width = 30 ft. How much fencing is needed?

SECTION 2 MULTIPLE CHOICE

Directions: Darken the letter of the best answer on the answer sheet. You may use a calculator for this section.

7. Examine the number line below. What best describes the value of the circle on the number line?

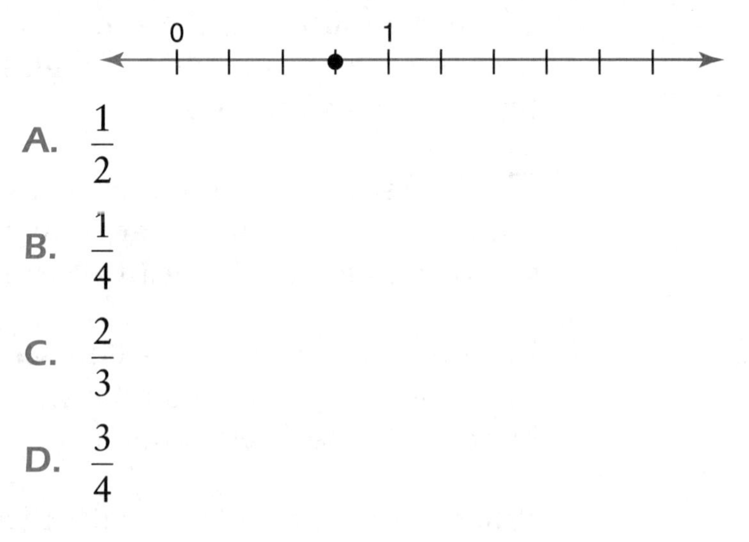

A. $\frac{1}{2}$

B. $\frac{1}{4}$

C. $\frac{2}{3}$

D. $\frac{3}{4}$

8. What is the prime factorization of 27?

A. $3 \times 3 \times 3$

B. $2 \times 2 \times 2$

C. 1×27

D. 3×9

9. Richardo bought 4 pounds of cheese. He saved $1.60 by using a coupon. How much did he save per pound of cheese?

A. $1.60

B. $1.00

C. 60¢

D. 40¢

10. Mr. Utley gave 10 students a mathematics exam. The scores were 79, 88, 100, 79, 84, 92, 68, 100, 68, and 68. What was the mode of these scores?

A. 68

B. 79

C. 92

D. 100

11. Which figure below has a perimeter of 16?

A. A square with side = 4

B. A rectangle with sides 8 and 2

C. A square with side = 8

D. A rectangle with sides 4 and 6

12. The table below shows the number of emails that Abe received each day for a week. What was the mean number of emails received by Abe?

Day	Emails
Monday	29
Tuesday	21
Wednesday	26
Thursday	24
Friday	30

A. 26

B. 27

C. 29

D. 30

13. In 2002, the state of Washington produced between 6,100,000,000 and 6,200,000,000 pounds of apples. Which could be the number of pounds of apples the state produced during 2002?

A. 6,110,000

B. 6,200,500,000

C. 6,100,578,000

D. 6,090,000,000

14. Which of the following statements about parallelogram *PQRS* appears to be true?

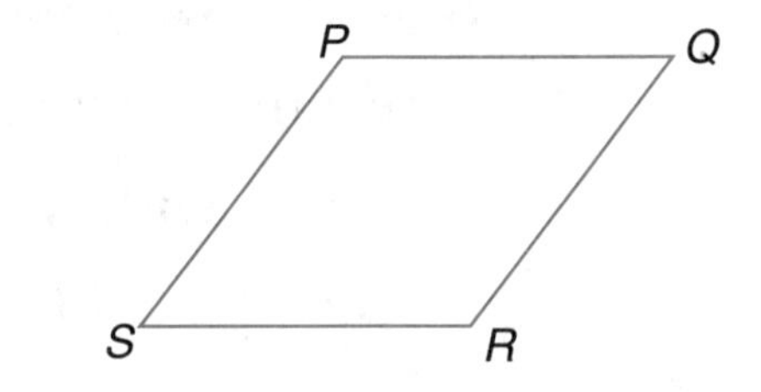

A. *PS* is parallel to *QR*.

B. *PS* is parallel to *SR*.

C. *PS* is perpendicular to *SR*.

D. *PS* is parallel to *PQ*.

15. Austin was buying pencils for school. He bought as many pencils as possible with $3.10. Pencils cost 10¢ each or $1.00 per dozen. How many pencils did Austin buy?

A. 30

B. 31

C. 36

D. 37

16. Which of the following is equal to 2 gallons?
   A. 6 quarts
   B. 8 quarts
   C. 20 cups
   D. 40 pints

17. Using his computer, Juan can type 20 words per minute. At this rate, how many words can Juan type in 2.5 minutes?
   A. 40
   B. 50
   C. 60
   D. 80

18. Lunch at Susan's school costs $3.25. She pays for lunch with a $5 bill. How much change does she receive?
   A. $1.25
   B. $1.75
   C. $3
   D. $3.25

EXTENDED CONSTRUCTED-RESPONSE QUESTIONS

Directions: Write your response in the space provided on the answer sheet. Answer the question as completly as possible. You may use a calculator for this section.

19. Monique worked four days at Hope's Fashion Shop. She earned $160.

- Make a table showing the total money Monique earns for days 1 through 5 and write a rule to determine the total money earned in n days.
- If the pattern continues, how much money would she earn if she worked a total of 20 days? Show work.

- If she worked 5 hours per day, how much does she make per hour? Show work and explain answer.

SECTION 3 MULTIPLE CHOICE

Directions: Darken the letter of the best answer on the answer sheet. You may use a calculator for this section.

20. Juan was on the Internet using Instant Messaging with two friends. He IMed with Paula for $1\frac{3}{4}$ hours, and to Michelle for $1\frac{1}{4}$ hours. How much time did Juan spend on the Internet?

    A. 2 hours
    B. 2.5 hours
    C. 3 hours
    D. 4 hours

21. Chase can throw a football $60\frac{1}{4}$ feet. Donovan can throw the same football 48 feet. How much farther can Chase throw the ball than Donovan?

    A. $108\frac{1}{4}$ feet
    B. $20\frac{1}{4}$ feet
    C. 20 feet
    D. $12\frac{1}{4}$ feet

22. Which of the answers below represents the product of b and 12?

    A. $12b$
    B. $12 + b$
    C. $b - 12$
    D. $12 - b$

23. Which of the following groups is composed of all equivalent fractions?

A. $\frac{1}{3}, \frac{2}{6}, \frac{3}{9}, \frac{4}{12}$

B. $\frac{1}{2}, \frac{1}{3}, \frac{1}{4}, \frac{1}{5}$

C. $\frac{1}{3}, \frac{2}{4}, \frac{3}{5}, \frac{4}{6}$

D. $\frac{1}{2}, \frac{1}{3}, \frac{2}{4}, \frac{2}{6}$

24. Sagan ran a 1-kilometer race. When he was $\frac{3}{4}$ of the way completed, how many meters did he have left to run?

A. 500

B. 300

C. 250

D. 100

25. Given a regular six-sided die, what is the probability of rolling a number greater than 5?

A. $\frac{1}{6}$

B. $\frac{2}{6}$

C. $\frac{3}{6}$

D. $\frac{4}{6}$

26. The square below is composed of 100 smaller squares. If each of the smaller squares is equal to 1/100, then what is the total value of the shaded squares?

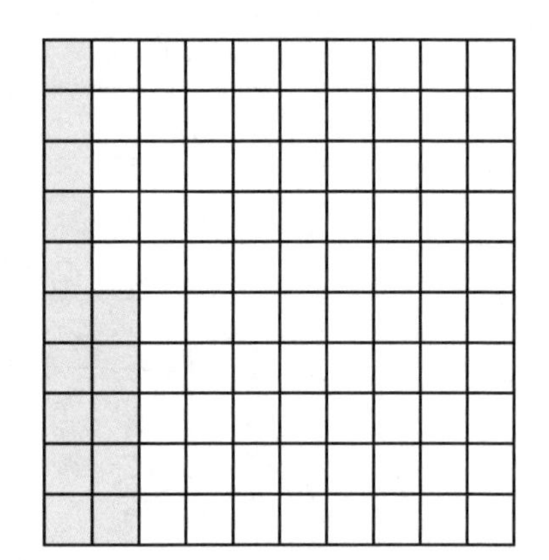

A. 15

B. 1.5

C. 0.15

D. 0.015

27. Karin is buying tile for her bathroom floor. The floor measures 10 feet by 15 feet. How many square feet of tile will she need to buy?

A. 50

B. 100

C. 150

D. 175

28. Jordyn is drawing a circle with a diameter of 6 inches. What is the radius of her circle?

A. 3

B. 6

C. 9

D. 12

29. Michelle and Michaela are ordering pizza for their class to sell during field day. They checked with 10 friends and found that 4 would buy a slice of pizza. Based on this, if 50 students will be attending, *about* how many slices of pizza will be bought?

A. 20

B. 40

C. 50

D. 100

30. What number is missing in the table below?

Input	Output
3	10
5	16
6	19
7	?

A. 20

B. 21

C. 22

D. 23

31. The carpet in Michelle's living room is 15 ft. by 20 ft. What is the perimeter of the room?

A. 35 feet

B. 70 feet

C. 140 feet

D. 300 feet

EXTENDED CONSTRUCTED-RESPONSE QUESTION

Directions: Write your response in the space provided on the answer sheet. Answer the question as completly as possible. You may use a calculator for this section.

32. John has four friends: Amanda, Brian, Charles, and Dwayne. He wants to pick two friends to play a game with him. Show how many ways he can pick two friends using a list, a chart, or a tree diagram.

SECTION 4 MULTIPLE CHOICE

Directions: Darken the letter of the best answer on the answer sheet. You may use a calculator for this section.

33. Eric has several coins in his pocket including four quarters, three dimes, two nickels, and one penny. If he reaches into his pocket and pulls out one coin, what is the probability that the coin will be worth less than a dime?

A. $\frac{7}{10}$

B. $\frac{5}{10}$

C. $\frac{3}{10}$

D. $\frac{2}{10}$

34. A triangle has two 45° interior angles. Which term correctly describes the triangle?

- **A.** It must be a right triangle.
- **B.** It must be an obtuse triangle.
- **C.** Its other angle is 45 degrees.
- **D.** It is an equilateral triangle.

35. The population of New Jersey on April 1, 2000, was 8,414,350. The place value of the 1 is

- **A.** millions
- **B.** hundred thousands
- **C.** ten thousands
- **D.** thousands

36. Which of the following shows the numbers in order from least to greatest?

- **A.** 1, 1.5, 1.6, 1.25
- **B.** 1, 1.25, 1.5, 1.6
- **C.** 1.6, 1.5, 1.25, 1
- **D.** 1, 1.6, 1.5, 1.25

37. The fifth grade class is collecting bottles for recycling. Each week, the class collects 78 bottles. At this rate, how many bottles will the class have collected at the end of 12 weeks?

- **A.** 936
- **B.** 956
- **C.** 976
- **D.** 1000

38. Kristen wants to plant her flowers with the same number in each row. How many flowers can be planted in 8 rows so that the same number of flowers are in each row?

A. 71

B. 72

C. 73

D. 74

39. Rebecca is 55 inches tall but Susan is only 49 inches tall. Which of the following correctly compares the height of each child?

A. $55 > 49$

B. $49 > 55$

C. $55 < 49$

D. $49 = 55$

40. Tyson has a bag of 30 marbles that contains 14 red marbles and 16 blue marbles. If Tyson reached into the bag without looking and picked one marble, what is the probability that he would pick a blue marble?

A. 1 out of 30

B. 14 out of 30

C. 16 out of 30

D. 10 out of 30

41. Using the digits 1–7 only once, what is the largest odd number you can make with a 6 in the thousands place?

A. 7,654,312

B. 7,546,321

C. 7,645,312

D. 7,546,231

42. What is the most reasonable estimate of the length of a city's swimming pool?

A. 2 meters

B. 30 meters

C. 5 kilometers

D. 200 kilometers

43. Given the number sentence $\square \div 6 = 7$. What number belongs in the $\square$?

A. 1

B. 13

C. 42

D. 84

EXTENDED CONSTRUCTED-RESPONSE QUESTION

Directions: Write your response in the space provided on the answer sheet. Answer the question as completly as possible. You may use a calculator for this section.

44. Two students measured the same length of their desks. Debbie said the measurement is 3. Tim said the measurement is 36. How can both students be correct? Explain your reasoning.

ANSWERS TO SAMPLE TEST #2

1. 30 (multiple of 3, divisible by 5, even, and less than 40)

2. 164

3. 9.72

4. 22

5. $(100 + 90 + 95 + 95)/4 = 380/4 = 95$

6. $2(20 + 30) = 100$ feet

7. **D.** $\frac{3}{4}$

8. **A.** $27 = 3 \times 3 \times 3$

9. **D.** $1.60/4 = 40¢

10. **A.** 68—appears the most (3 times)

11. **A.** Square with side 4 has perimeter $4 \times 4 = 16$

12. **A.** $(29 + 21 + 26 + 24 + 30)/5 = 26$

13. **C.** 6,100,578,000

14. **A.** *PS* is parallel to *QR*

15. **D.** 37. One dozen = $1, 3 dozen = $3, 3 dozen + 1 = $3.10

16. **B.** 8 quarts

17. **B.** $50 = 20 \times 2.5$

18. **B.** $1.75 = $5 – $3.25

19. You will receive full credit if:

- You correctly find that the rule is $d \times 40$ and fill in the table with the answers below.

Days, d	Money Earned, m
1	40
2	80
3	120
4	160
5	200
d	$d \times 40$

- You correctly find that Monique earns \$800 in 20 days (20 days × \$40 per day = \$800).
- You correctly find and support findings that Monique earns \$8.00 per hour (40 ÷ 5 = 8).

20. C. 3 hours = 1.75 + 1.25

21. B. 60.25 − 48 = 20.25

22. A. 12b

23. A. $\frac{1}{3}, \frac{2}{6}, \frac{3}{9}, \frac{4}{12}$

24. C. He ran $\frac{3}{4}$ of 1,000 = 750 meters. He has 250 meters left to run.

25. A. $\frac{1}{6}$

26. C. 0.15 = 15/100

27. C. 10 × 15 = 150 sq ft

28. A. $\frac{6}{2} = 3$

29. **A.** $(4/10) \times 50 = 20$

30. **C.** 22 (Output = $3 \times$ Input + 1)

31. **B.** 70, perimeter = $2\ (15 + 20)$

32. You will receive full credit if:

- You correctly find there are 6 ways to pick two friends (AB, AC, AD, BC, BD, and CD).
- You show all work and explain how you got your answer.

33. **C.** $\frac{3}{10}$ (3 coins are less than 10 cents)

34. **A.** If two angles are each 45°, the third angle must equal 90°.

35. **C.** ten thousands

36. **B.** 1, 1.25, 1.5, 1.6

37. **A.** $78 \times 12 = 936$

38. **B.** 72 is divisible by 8

39. **A.** $55 > 49$

40. **C.** 16 out of 30

41. **B.** 7,546,321

42. **B.** 30 meters

43. **C.** 42 ($42 \div 6 = 7$)

44. You will receive full credit if:

- You correctly state that Debbie is referring to 3 feet and Tim is using 36 inches and show that 3 ft = 36 inches.

INDEX